D1402349

THE TELEPHONE

THE TELEPHONE

THE LIFE STORY
OF A TECHNOLOGY

David Mercer

GREENWOOD TECHNOGRAPHIES

GREENWOOD PRESS
Westport, Connecticut • London

LCC SOUTH LIBRARY

Library of Congress Cataloging-in-Publication Data

Mercer, David, Ph.D.
   The telephone : the life story of a technology / David Mercer.
      p.   cm. — (Greenwood technographies, ISSN 1549-7321)
   Includes bibliographical references and index.
   ISBN 0–313–33207–X (alk. paper)
   1. Telephone—History.   2. Telegraph—History.   3. Cellular telephones—History.
   4. Telephone—Social aspects.   I. Title.
   TK6015.M47   2006
   621.38509—dc22   2006021452

British Library Cataloguing in Publication Data is available

Copyright © 2006 by Greenwood Publishing Group

All rights reserved. No portion of this book may be
reproduced, by any process or technique, without the
express written consent of the publisher.

Library of Congress Catalog Card Number: 2006021452
ISBN: 0–313–33207–X
ISSN: 1549–7321

First published in 2006

Greenwood Press, 88 Post Road West, Westport, CT 06881
An imprint of Greenwood Publishing Group, Inc.
www.greenwood.com

Printed in the United States of America

The paper used in this book complies with the
Permanent Paper Standard issued by the National
Information Standards Organization (Z39.48–1984).

10 9 8 7 6 5 4 3 2 1

TK
6015
.M47
2006

APR 0 4 2008

Contents

Series Foreword

In today's world, technology plays an integral role in the daily lives of people of all ages. It affects where we live, how we work, how we interact with each other, what we aspire to accomplish. To help students and the general public better understand how technology and society interact, Greenwood has developed *Greenwood Technographies*, a series of short, accessible books that trace the histories of these technologies while documenting *how* these technologies have become so vital to our lives.

Each volume of the *Greenwood Technographies* series tells the biography or "life story" of a particularly important technology. Each "life story" traces the technology from its "ancestors" (or antecedent technologies), through its early years (either its invention or development) and its rise to prominence, to its final decline, obsolescence, or ubiquity. Just as a good biography combines an analysis of an individual's personal life with a description of the subject's impact on the broader world, each volume in the *Greenwood Technographies* series combines a discussion of technical developments with a description of the technology's effect on the broader fabric of society and culture—and vice versa. The technologies covered in the series run the gamut from those that have been around for centuries— firearms and the printed book, for example—to recent inventions that have rapidly taken over the modern world, such as electronics and the computer.

While the emphasis is on a factual discussion of the development of the technology, these books are also fun to read. The history of technology is full of fascinating tales that both entertain and illuminate. The authors—all experts in their fields—make the life story of technology come alive, while also providing readers with a profound understanding of the relationship of science, technology, and society.

Acknowledgments

I am indebted to the following people who provided the collegial support which helped make the writing of the manuscript possible: June Aspley, Kerry Ross, and Robert Brown. I would like to dedicate this book to Shirley Mercer, who, for family and friends alike, has never been more than a phone call away, and Ella and Zoe who hold the future in their hands.

Introduction

The life of the telephone can be thought of as consisting of three phases, the telegraph, the standard telephone, and the mobile. It is important to note that these phases intersect and overlap rather than follow a pattern of simple beginnings and ends. There are, as can be expected, a number of places where events and technological developments don't fit exactly into neat periods. For example, while it is correct to state that the telephone was invented in 1876, the year of Alexander Graham Bell's famous patent, it's important to remember it took a number of years for the telephone to take on the form and meaning with which most readers are familiar. The "early" telephone coexisted with a vigorous and successful telegraph industry. Its invention was inspired by trying to improve the electric telegraph, and in its early days it was often described as the "speaking telegraph." Similarly, aside from the mobile telephone's massive recent growth, the "standard telephone" is still integral to "day-to-day" life for the majority of people. To further complicate a neat chronological narrative the mobile telephone's "conception" can be traced back to the late 1940s with attempts to link radio and telephony although it would really only be in the 1990s that the mobile telephone (cellphone or handy) would become a mass consumer product.

Keeping these challenges in mind the book is organized into nine chapters tracing in chronological order the three phases of the life story of the telephone.

Chapter 1 traces the period from 1780 to 1870 and documents the invention and development of the telegraph. It starts with the first visual telegraph systems developed by the Chappe brothers in France around the time of the French Revolution. The Chappe telegraph became one of the most significant technologies of Napoleonic France. The State used it to coordinate warfare and maintain political control. It represented one of history's first extensive technological communication systems and inspired a range of new ways of thinking about information, codes, and the capacity for, and implications of, controlling time and space. The chapter moves on to trace the origins of the electric telegraph. From the early part of the nineteenth century there were numerous scientists preoccupied with electricity, and as the century moved on there were numerous inventors fascinated with the potential practical uses for electricity. In 1837 Cooke and Wheatstone patented and developed the first electric telegraph system in Britain; they were to be followed soon after in the United States by Morse in 1841. Morse's system was technologically more simple but relied on users learning a code, and it ultimately dominated telegraphy. The electric telegraph became more than just a vehicle of the State and also took off as a business tool, allowing the coordination of goods, commerce, and news. At the time of its emergence the electrical telegraph captured public and intellectual imagination. There was an awareness that, as now messages could travel massive distances almost instantaneously, time and space would no longer be the same. The telegraph appeared in popular culture in many guises: as a representation of "the nervous system" of the nation, as evidence for spiritualism (if communication of "intelligence" through time and space by electricity was possible, why not then, the possibility of communication with the "after life"), and, based on the assumption that one of the main sources of warfare was simply poor communication, as a tool for encouraging world peace. The pinnacle of the age of the telegraph was the laying of the transatlantic cable that allowed messages to be sent instantaneously (in theory, at least) between London and New York. Many viewed it as the major technological triumph of the nineteenth century.

Chapter 2 focuses in some detail on the single most important year in the life of the telephone, the year of its "birth," 1876. It notes the way the telephone, or "speaking telegraph" as it was sometimes known then, emerged from attempts to build "harmonic telegraph" systems that would allow multiple messages to be sent down the same telegraph line simultaneously. The chapter also traces the controversy over whether or not Alexander Graham Bell should be given credit as the inventor of the telephone or whether this honor should be granted to his rival Elisha Gray.

Chapter 3 starts in 1876, the year of the telephone's invention, and traces events uptil 1893, the year that Bell's original telephone patents expired. It outlines the early struggles to promote the telephone and find uses for it. To many at the time it wasn't immediately clear what advantages speaking over the telegraph offered over sending text. This was also a period where a number of inventors such as Thomas Edison struggled to improve the telephone and disputed Bell's rights to its patent. This era also featured the development of the first business models for organizing telephony and the emergence of the first Bell company telephone monopoly.

Chapter 4 begins by describing a short-lived period of competition starting in 1893 in which the Bell telephone monopoly was briefly under challenge from a variety of independent companies. It goes on to outline the reconsolidation of the Bell monopoly under the second term of its general manager Alfred Theodore Vail. During this time Vail helped lay the foundations for the Bell telephone system, which would rely on integrating local and long-distance services, and strike deals with government regulators by providing investments in new technology and expansion of services, in return for government protection from competition. This was an era where Bell employed large numbers of women switchboard operators in preference to developing automatic switches. The highlight of this period in telephone technology is the triumph of the opening of the transcontinental telephone line from New York to San Francisco in 1915. The chapter closes by documenting the much more rapid diffusion of telephones during this time in the United States than in other countries.

Chapter 5 continues to map the consolidation of the Bell telephone system between the world wars (1918–1945), and pays special attention to the shift from the promotion of the telephone near the beginning of this era as a business tool to its promotion as a means for enhancing sociability and "day-to-day" conversation.

Chapter 6 traces the "calm before the storm" between 1945 and the 1970s. This post–Second World War period marked the birth of numerous important technologies such as the transistor, information theory, radar, microwaves, fibreoptics, and computers—technologies that would ultimately be associated with emergence in the 1970s and 1980s of the so-called "information society." While these radical technological changes were taking place, the standard telephone, from the point of view of the user, enjoyed a long period of stability, becoming a "taken-for-granted" part of life. This period of stability is used as an opportunity to attempt the difficult task of providing a general account of the social impacts of the "standard" telephone.

Chapter 7 notes the turbulence of the deregulation of telecommunications and the new microelectronics emerging from the late 1970s and early 1980s—a turbulence that continues till the present. The way the telephone fits into the current age of digital convergence of the "information society" is also traced.

Chapter 8 discusses the birth of the mobile telephone, paying particular attention to the period of the 1980s to the present. In this chapter the life story of the telephone shifts focus away from the activities of Bell in the United States, the main site of activity for the development of the "traditional telephone," to map the mobile telephone as a global phenomenon. A brief comparison of the way the policies of different nations have shaped the early development of the mobile is provided.

In Chapter 9 discussion focuses mainly on the period from the 1990s to the present and explores some of important social impacts of the mobile telephone. This chapter revisits some of the themes covered in Chapter 6, where the social impacts of the standard telephone are discussed, such as whether or not using the mobile telephone leads to shallower or deeper patterns of communication.

As far as possible the book avoids using technical jargon, but it is inevitable that in writing the history of technology some technical issues need to be documented. As far as possible technical matters are explained as they arise within the text and a number of "line drawings" have been provided to explain the principles relevant to the Chappe telegraph; electric telegraph; telephone switching; the standard telephone; the cellular telephone system; and the electromagnetic spectrum. A brief glossary explaining technical terms and acronyms also appears at the end of the text. The chronological narrative is as comprehensive as possible but it is inevitable that not every aspect of the development of the telephone can be covered in depth and the enthusiastic reader may find that various issues can be traced further by referring to the sources in the Bibliography.

While the text does not explicitly engage in theoretical debates from the academic fields of science and technology studies or the social history of technology, an effort has been made to embody the insights from these areas of scholarship into the narrative. The reader will come to see that the life story of the telephone has not merely unfolded along some kind of inevitable technological trajectory, but rather that scientists, inventors, designers, business managers, workers, and users have all played their part and that the life story of the telephone is also their story.

Timeline

1791	On March 2 Claude and Rene Chappe perform a demonstration of a visual telegraph system to an audience of local government officials. One of the local officials, Miot de Mileto, suggests the name telegraph or "far writer."
1791–1793	Claude Chappe develops a more refined codebook and a signaling device that uses arms and pullies.
1794	The first major telegraph line is built in May, from Paris to Lille.
1795	The British Admiralty orders construction of towers between London and ports of England's South Coast. This British visual telegraph uses a system of opening and closing shutters.
1797	Encyclopedia Britannica suggests that better communication offered by the telegraph will help create possibilities for a more peaceful society.
1800	Improvements in developments in the technologies of batteries by scientists such as Alessandro Volta helps produce more reliable sources of electricity that make experiments with the telegraph easier.
1816	Telegraph system built in England in by Francis Ronalds.

1820	Hans Christian Oersted in Denmark notes that electricity passing through a wire has an effect on compass needles because it produces a magnetic field.
1820s–1830s	Joseph Henry in the United States performs experiments with electromagnets and telegraph devices.
1835	Morse takes an appointment as Professor of the Literature of the Arts of Design at New York University.
1837	Cooke and Wheatstone secure a patent in England for a needle electric telegraph.
	Morse decides to try to build his own electric telegraph and employs the scientific assistance of Leonard Gale and the mechanical assistance of Alfred Vail. Vail contributes significantly to the development of Morse code.
	American physicist William Charles Page explores the possibilities of producing sounds by rapidly magnetizing and demagnetizing metal rods.
1840	Morse obtains a patent for the electric telegraph in the United States.
	France has over 3,000 miles of visual telegraphs tied together by line of sight from over 500 towers.
1844	The first official Morse telegraph message is sent from Baltimore to Washington, setting the tone for the occasion. Morse's mundane dots and dashes spell out the portentous words "What hath God wrought."
1845	Cooke secures a lucrative contract with the British Admiralty to build an 88-mile long electric telegraph line between Portsmouth and London.
1850	Britain has 2,215 miles of telegraph wire. This early development is strongly associated with the growth of railways.
	The United States has 12,000 miles of telegraph wire in 1850, initiating a period of rapid growth (by 1852, 22,000 miles, and by 1854, over 30,000 miles of line).
1851	Morse's system becomes the European standard.
	The first undersea telegraph cable is constructed between England and France.
1854	Belgian-born inventor Charles Bourseul presents a report to the *Academie Des Sciences*, which discusses the possibility of transmitting sound vibrations via electricity.

1857   Cyrus Field creates the Transatlantic Telegraph Company.

The first attempt at laying the transatlantic cable starts out from Valentia Island in Ireland in August 1857 but is a failure marred by repeated breakages of the cable.

1858   In June, another attempt to lay the transatlantic cable fails. Expedition setting out in July 1858 finally achieves the "successful" laying of the transatlantic cable. On the August 5 the cable is "landed" and, on August 17 the first official transatlantic telegraph message is transmitted from Queen Victoria to President James Buchanan. Within a month the cable no longer works.

1861   The first demonstrations of an actual "telephone-like" device are performed in Germany by Johan Philip Reis at the Physical Society of Frankfurt.

1865   Attempts to lay a new improved transatlantic cable designed with the assistance of Scottish Physicist William Thomson. On June 24 the world's largest ship *The Great Eastern* sets out on its first attempt to lay transatlantic cable, two-thirds of the way across the Atlantic. The cable snaps.

1866   On Friday, July 13, another attempt to lay the transatlantic cable is made, but this time it is a success. *The Great Eastern* not only lays the new telegraph line but a month later recovers and repairs the lost cable of 1865. There are now two "working" transatlantic cables.

Western Union, specializing in short business telegraph messaging, becomes the first U.S. firm to span the whole continent.

1871   On June 10, Samuel Morse is publicly hailed as the father of the telegraph with the unveiling of a bronze statue in Central Park, New York.

1875   Alexander Graham Bell visits Joseph Henry at the Smithsonian, gains financial support from Gardiner Green Hubbard and George Sanders, employs an assistant, Thomas Watson, and begins working with a variety of experimental telephone-like harmonic telegraph devices. At the same time Elisha Gray and a number of other inventors are also doing similar work.

1876   On February 14, Alexander Graham Bell files a patent for "improvements to the telegraph, and the magneto-electric telephone, including a speaking telegraph." Elisha Gray files a Patent Caveat for a "talking telegraph" a couple of hours later.

Bell's patent is formally issued on March 3, 1876. This patent U.S. Number 174,465 is possibly the most financially valuable ever issued.

Bell gives public displays of the telephone.

In late 1876 Bell's financial backers Hubbard and Sanders unsuccessfully attempt to sell their telephone patent rights to Western Union for $100,000.

| 1877 | On July 9, Hubbard, Bell, and Sanders form the Bell Telephone Company. |

In December 1877 Western Union creates the American Speaking Telephone Company.

Western Union begins to adopt telephones designed by Edison, Gray, and others.

| 1878 | Edison and Berliner develop the idea of contact pressure transmitters that improve the clarity and strength of telephone transmission, to be further improved by Francis Blake. |

Further telephone innovations of 1878 included Thomas Watson's telephone ringer and the establishment of the first telephone exchanges.

The first telephone is installed in the White House, for President Rutherford B. Hayes.

| 1878–1887 | Theodore N. Vail's first term as general manager and president of *Bell*. |

Eighteen years of litigation, which will see *Bell's* patents being tested in 600 separate cases, begins.

| 1879 | *Bell* and Western Union settle their patent disputes (although various claims by rival inventors linger on). |

United Telephone Company formed in the United Kingdom.

| 1879–1898 | More than 86 new automatic switching systems are patented and offered for sale to *Bell*. These devices are not to be widely used for a number of decades as *Bell* persists with using female telephone operators instead. |

| 1893–1894 | *Bell's* key telephone patents expire, and there is a burst of activity as new companies enter into the telephone business. This era of competition is marked by a significant cut in the cost of telephones, and also revenues per phone for *Bell*. |

| 1899 | Almon Strowger patents an automatic telephone switching system. Its basic design will influence design of telephone switching technology well into the twentieth century. |

| 1900 | Michael Pupin, a professor of Electromechanics at Columbia University, takes out a patent for the "loading coil." Similar research |

is being conducted by George Campbell for *Bell*. The loading coil is used to help amplify signals for long-distance telephone lines.

1906	The "audion" is invented by Lee de Forest. Harold Arnold, an AT&T PhD researcher, applies "new theories of electromagnetism" for adapting the audion to the needs of the telephone and helps develop the "high vacuum thermionic tube."
1910	Formal advertising for telephones commences. It is geared toward businessmen and emphasizes the role of the telephone in saving time, planning, impressing customers, being modern, and keeping in touch with work while on vacation.
1907–1919	Vail's second term. With the financial support of the Banker J. P. Morgan numerous independent phone companies are bought up and integrated into the Bell system.
1908	Vail begins advertising nationally the slogan that would later become famous: "One System, One Policy, Universal Service."
	Vail appoints J. Carty and takes considerable interest in research and development of telephone technology, planting the seeds for the development of the Bell Laboratories.
1911–1912	In the United Kingdom, the GPO (General Post Office) takes over most of Great Britain's telephone services and then refuses to grant new licenses after December 31, 1911. It then finally takes over telephony completely in 1912. Most telephone systems that emerge across the world follow similar patterns of state ownership.
	Vail's strategies prove successful, and by 1912, 83 percent of independent telephone companies are connected to *Bell's* wires.
1913	The U.S. Justice Department informs Vail that the Bell system is bordering on breaching the Sherman Antitrust Act. Rather than risk further antagonism from government authorities or litigation, Vail strategically compromises in a number of key areas, signing off the Kingsbury Commitment of 1913 (drafted by an AT&T vice president, Nathan Kingsbury).
1914	There are 1.7 telephones per 100 persons in the United Kingdom compared to 9.7 in the United States.
1915	On January 25 the 4,300 mile long transcontinental telephone line is opened. Vail puts considerable energy into publicizing its triumphant opening.
1920	Radio technology emerges as commercially important, the major players AT&T, General Electric, and the Radio Corporation of America sign a cross licensing agreement, covering 1,200 patents (Westinghouse also enters the agreement in 1921). The parties agree

to grant rights to the others to use patents but limit the markets each party could apply the new technology to. *Bell* agrees not to enter into the actual business of radio broadcasting in return for maintaining exclusive control over public markets for radiotelephony and its existing wires.

Bell becomes the first company in the United States to generate $1 billion in revenue.

1921	The Willis–Graham Act cements the rationale of the Kingsbury agreement in law allowing *Bell* to be exempt from antitrust limits on purchasing telephone companies.
1925	Bell Laboratories open.
1929	About 42 percent of all U.S. households have telephones, a figure which dips during the Depression to 31 percent, to then reconsolidate to 37 percent by 1940.
1930s	Telephone companies begin to acknowledge in their advertising the role of telephones for "day-to-day" sociability and not just business.
1937	*Bell* releases the Henry Dreyfus designed Bell "300" telephone (the model T of telephone design).
1938	The Walker Report describes the pattern of regulation surrounding the Bell system as unworkable. The first crossbar switching systems come into use.
1940s	Coaxial cables are developed. These cables offer much better insulation allowing a greater range of frequencies to be transmitted and in turn a much greater quantity of information to be carried. They become important for improving long-line service and television transmission.
1944	*Bell* controls 83 percent of all U.S. telephones, 98 percent of all long-distance wires, and is the world's largest firm with $5 billion assets.
1945	The Second World War encourages significant technological developments such as radar, microwave technology, and early electronic computers.
1947	*Bell* offers limited mobile radiotelephone highway service between New York and Boston, operating from automobiles. Technology enabling the radiofrequency spectrum to be divided to serve large numbers of users is in its infancy. So there are limits to the number of users the system can serve.

D. H. Ring and W. R. Young begin to develop the principles of cellular communication, a way of dividing the radiofrequency spectrum to avoid interference that will allow for a greater number of signals/user.

1948   On July 1, Bell Laboratories reveals one of the most important technologies of the twentieth century: the Transistor. This device is the joint invention of William Shockley, John Bardeen, and Walter Brattain.

Claude Shannon of the Bell Laboratories publishes *The Mathematical Theory of Communication*. Shannon's work encourages the development of information theory, which in turn later contributes to the development of computers and the Internet.

1950   62 percent U.S. residences hold telephones subscriptions.

1950s   Firms wishing to enter into microwave broadcasting continue to challenge *Bell's* protected position. There is steady lobbying for firms to be able to operate private microwave systems.

A basic system of mobile radiophones begins operation in Sweden.

1962   The Telstar satellite designed by the Bell Laboratories is launched. 80 percent of U.S. residences hold telephone subscriptions.

1964   *Bell* displays at New York World Fair a model of the Picturephone. It proves to be a commercial failure.

1965   The stored program telephone switch system is first put into commercial use after approximately 30 years development and $500 million of investment.

1967   The chief engineer of Swedish Telecom Radio, Carl-Gosta Asdal, suggests that Sweden should develop an automated mobile telephone network integrated with the landline network.

1968   *Carterphone* case: Texan entrepreneur wins the legal right for customers to attach Carterphone instruments to AT&T's lines.

1969   Nordic countries—Denmark, Norway, and Finland—form the Nordic Mobile Telephone Group.

1970   90 percent of U.S. residences hold telephone subscriptions.

1974   In an act symbolically marking the beginning of the end of the traditional organization of the telephone system, the U.S. Justice Department files an antitrust suit revisiting their long-standing concerns that it is inappropriate for AT&T and Western Electric to be part of the same company, the Bell system.

1976   *Comstar* satellite is launched. It carries up to 30,000 calls at the same time.

1978   The first U.S. cellular telephone system is set up by *Bell*. It has a capacity of 2,000 users who can link up via telephones carried in cars to base stations and the traditional telephone system.

1970s–1980s The first fiberoptic cables are developed by Corning Glass.

            Take-off of microelectronics and computer industries.

            Social theorists such as Daniel Bell start writing that a Post-Industrial or Information Society is emerging with communications and information technologies replacing the role of traditional industry.

            Large businesses are becoming increasingly dependent, for coordinating things like cash flows, investments, and production, on the rapid flow of huge amounts of digital information that passes through telephone lines. This encourages the development of modems and PABXs (Private Automatic Branch Exchanges) and other mainly business-oriented telephone technologies.

            From the late 1970s the economic policies of Margaret Thatcher and then Ronald Reagan spark intense and ideologically charged debates about the appropriate role for economic regulation across much of the Western World.

1979   In January, the *Bell System Technical Journal* devotes an entire issue to cell phones but shows little interest in pursuing their immediate development.

1981   The NMT (Nordic Mobile Telephone) system is launched.

1982   Divestiture of the Bell system. On January 8 an agreement is reached for the Bell system to be split up.

1983   AT&T has revenues of $65 billion, 1 million employees, assets worth $150 billion, and 84 million customers.

1984   The 1982 agreement for the divestiture of *Bell* is legally consented to on January 1. AT&T retains control of Western Electric and is allowed to maintain an interest in long-distance operations but must divest itself of its local operating companies. These local operations are taken over by the seven so-called Baby *Bells*.

1980s   Surveys suggest three-fourths of all local calls are made for social reasons to family and friends. Another survey suggests almost 50 percent talk on the telephone to friends or relatives every day.

            The Divestiture of *Bell* also holds relevance for the way the telephone industry is organized across the world, stirring debates (still resonating in many countries till the present) about the

	appropriate mix of regulation vs. deregulation of telecommunications.
1982	Meetings are held in Stockholm between engineers and administrators from 11 European countries. It considers the development of a Europe-wide so-called GSM mobile phone system. This system ultimately becomes popular worldwide and marks the birth of the "second generation" of mobile telephones.
1984	Motorola releases its first commercial mobile telephone with a suggested price range of $3,000 to $4,000.
1987	Approx 2 percent of the population of Nordic countries are mobile telephone subscribers.
	"State of the art" mobiles are between 700–800 grams or about 2 pounds in weight.
1980s–1990s	Governments and business push to develop a fully digital information infrastructure described under the acronym of ISDN (Integrated Service Digital Network). More variety of services to consumers and businesses are provided but questions over service quality "regulation vs. deregulation" and "universal service" remain.
1990s to the Present	Global burst in popularity of mobile telephones and the Internet raise questions about the future role of traditional telephone.
	"Texting" becomes a huge global phenomenon. It becomes extremely popular in Southeast Asian countries (such as Singapore and the Philippines) with Europe, China, and Australia following closely behind. Texting is initially less popular in the United States.
1993	Launch of the Japanese digital cell phone system.
1999	In Norway 80 percent of 13- to 20-year-olds own a mobile phone.
2000 to the present	Introduction of the "third generation" (3G) of mobile telephones begins. 3G is based on the idea that mobile telephones should be able to integrate with and even substitute for the functions performed by personal computers.
2000	2 billion mobile users estimated worldwide.
2001	In the United Kingdom, 90 percent of people under 16 own a mobile phone.
2002	United States 13 percent of 12- to 14-year-olds own a mobile.
	One of Europe's largest telephone operators Orange announces on April 9 that 750,000 of its prepaying customers have neither received nor made a call in the last 3 months.

2003   *Newman v. Motorola.* Christopher Newman, a Baltimore
        Neurologist, claims his use of a mobile phone caused the
        development of a brain tumor behind his right ear and attempts to
        sue Motorola. Dozens of other cases pending against cell phone
        manufacturers at the time, and estimates of possible liabilities in
        excess of $6 billion. The case is heard in the United States District
        Court for Maryland where Newman's claims are dismissed.

2004   In United States 40 percent 12- to 14-year-olds own a mobile.

2005   Nokia files patent for Morse-code cell phones.

        About 16 million teens and younger have cell phones in the United
        States.

2006   Predictions that 50 percent of the world's population will be using a
        mobile phone by the end of 2009.

1

The Invention and Development of the Telegraph, 1780s–1870s

◆

The story of the telephone begins with the telegraph. It is no accident that the term "telegraph pole" is still used today to describe the tall wooden posts that, in many places, carry the wires for telephones. It would be from changes in the way communication was thought of, the possibility of transmitting information via electricity, and attempts to improve the telegraph, that the telephone would later emerge. The telegraph was perceived as a radical technology in its own time. By the mid-nineteenth century in Great Britain it was spoken of as the nervous system of the Nation and the Empire. The networks of cables were perceived as nerves and the telegraph office the brain that ordered, received, and transmitted messages. This metaphor worked two ways as doctors and physiologists in return drew inspiration from the telegraph to help explain the human nervous system (Rhys-Morus 2000, 458). The telegraph influenced the ways in which people viewed time and space and attracted utopian speculations that improved communication would encourage world peace.

Prior to the telegraph, physical space and time constituted much more obvious limits to communication. Books and letters captured ideas in forms where they could be reproduced, preserved, and communicated to others, but this took time, effort, and literacy—a message could be carried by a runner or by someone on horseback, and more exotic options such as carrier pigeons, talking drums, and smoke signals were also possible. The distance the message needed to be carried made a critical difference to how long it took to be conveyed. For the governments of the competitive, militaristic, and increasingly bureaucratic European nation states of the late eighteenth and early nineteenth centuries, breaking away from these constraints had obvious value. Prompt news from a battlefront, or of political dissent in the provinces, allowed faster, more critical responses and opportunities for stronger political control. It was from attempts to solve these problems of coordination and control that the telegraph would first emerge in postrevolutionary France at the end of the eighteenth century.

THE CHAPPE (VISUAL) TELEGRAPH

The pioneers of the telegraph were the Chappe brothers in France in the 1780s. Their first experiments used systems of clocks, sounds, and codes. The brothers would stand as far apart as they could while still being in earshot. A loud clanging noise was then used to synchronize the clocks each brother held. Further clanging noises were made from one brother to the next as the second hand of the clocks passed over chosen numbers on their clock faces. A code was devised so that these numbers would correspond to letters, words, and phrases. This system was limited by the distance a sound could be transmitted. So the Chappes quickly moved on to develop systems which relied on visual signals that, with the benefit of observers with telescopes, could be seen at much greater distances. On March 2, 1791, Claude and Rene Chappe performed a demonstration of their system to an audience of local government officials. They transmitted a message between a castle in Brulon and a house in Parce, a distance of 10 miles. Their demonstration involved the use of black-and-white visual panels, synchronized clocks, and telescopes. Numbers were placed on the clock face which, in turn, corresponded to words and phrases recorded in a codebook. As the second hand moved over a certain number, the panels of one operator were flipped from black to white. Using a telescope, another operator, at a distance, could record which number was being referred to on a clock synchronized with the first operator. Their demonstration was successful. One of the local officials, Miot de Mileto, suggested the name telegraph or "far writer" (Standage 1998, 9–11).

Between 1791 and 1793, Claude Chappe developed a more refined codebook and signaling device. This involved a system of small and longer arms that could be rotated into different positions. One arm represented a page in a codebook, another a word on the page of a codebook: 8,836 words and phrases could be represented. One operator worked with a smaller version of the signaling device that was attached via pullies to a scaled-up version that could operate from the top of a tower. If numerous towers were constructed in a line of sight, approximately every 10 miles, messages could be transmitted visually over long distances in quick time. Chappe gained support of the French government (the National Convention). This resulted in the first major telegraph line being built in May 1794 from Paris to Lille (Standage 1998, 9–14). The development of the visual telegraph was strongly tied to military concerns; one of the champions of the expansion of the telegraph would be the famous general Napoleon Bonaparte who came to power in France in 1799. The telegraph was also viewed as a device through which central authorities could maintain social control. From the outset there was resistance to anything other than government ownership and operation of the system. The government went as far as staffing intermediate towers with workers who were deaf (human relays employed in the Chappe system were called "mutes") to maintain security (John 1998, 195). By 1832, Abraham Chappe, Claude's younger brother would praise the telegraph as a tool to "carry to the centre of government, at the speed of thought, all political feeling . . . it gives more unity of action . . . when the government has to be ready to defend itself against attacks, when each minute must be efficiently used" (quoted in Flichy 1995, 18).

The idea of telegraphic communication also had various subtle links with changing perceptions of space and time. All these concerns featured prominently in France in the late eighteenth and early nineteenth centuries: a time of the Enlightenment, the French Revolution, and the New Republic. On August 17, 1794, one of Chappe's contemporaries, a member of the Committee for Public Safety, declared to the National Convention, "[b]y this invention [Chappe telegraph] distances between places disappear in a sense" (quoted in Flichy 1995, 9). Promoters of the telegraph also suggested it was a worthy replacement in the popular imagination for the cathedral bell towers which symbolized the traditional influence of the church in French life, and interest in telegraphic codes was also linked to calls for new forms of measurement, universal languages, and a new calendar.

The general features of the Chappe design were copied in other European countries, such as Britain and Sweden, and used primarily, as in France, to assist government in warfare and security. In 1795, the British Admiralty ordered construction of towers between London and the ports of England's south coast. Some commercial purposes were also pursued to assist merchants in coordinating shipping around ports such as Liverpool, Southampton, and London. This British visual telegraph used a system of opening and closing shutters rather than the arms of the Chappe system (Standage 1998, 16–18).

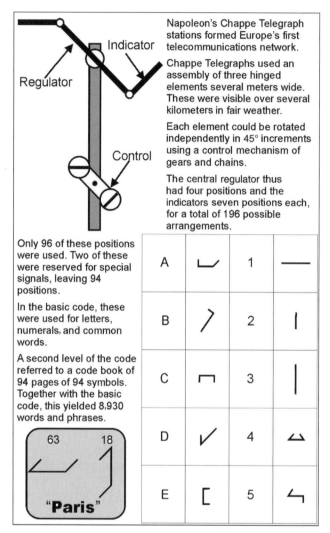

The Chappe Telegraph, Europe's first telecommunications network. Courtesy Robert B.K. Brown, 2006.

While the visual telegraph relied on fairly basic technological hardware, it still represented the materialization of a number of key ideas that would influence the development of later communication technologies. The system contained four key innovative features:

1. Although not instantaneous, speeds of transmission of information were much faster than preceding technologies, and increasing the speed of transmission of information was a key aim;

2. The system represented the establishment of permanent communications networks that could be steadily further expanded over time;
3. Specialized bodies emerged that oversaw the operation and development of communications; and
4. The system encouraged the development of theories about transmitting information such as codes and universal languages, not just the message (Flichy 1995, 31–32).

New ideas about organizing space, time, and information were also materialized by improvements in roads and shipping, and in the United States, in particular, the expansion of government postal services (John 1998).

Despite the success of the visual telegraph, and the fact that the Chappe name has been passed down through history, Claude Chappe did not fare so well in his own lifetime. After his early successes he proposed further ambitious systems and modifications of his designs. He nevertheless became increasingly paranoid, depressed, and hurt by criticisms made by rival inventors. On January 23, 1805, he committed suicide by jumping into a well outside the telegraph administration building in Paris. On his tombstone is engraved a telegraph tower indicating the signs for "at rest" (Standage 1998, 18). After Claude's death, his family was left to continue the work of lobbying for the expansion of the visual telegraph system in France.

THE ELECTRIC TELEGRAPH

Despite the considerable success of the visual telegraph (France was estimated to have over 3,000 miles of visual telegraphs tied together by line of sight from over 500 towers by the 1840s), the system had obvious drawbacks: it relied on a large number of skilled operators, it was expensive to operate, and it was limited by fog, rain, darkness, and unsuitable terrain. These constraints, coupled with new ideas about the possibilities of using electricity for communication, would ultimately lead to the visual telegraph's replacement. Understanding how electricity worked fascinated eighteenth- and nineteenth-century "natural philosophers," the forerunners to modern scientists, and an emerging group of inventors.

Like many important inventions, the electric telegraph had many parents, and in its infancy took many forms. Ideas for an electric telegraph system can be traced back to the same period as the early growth of the visual telegraphic system. For instance, between 1753 and 1837 at least 60 experimental electrical telegraph systems were constructed. This included one notable system, built in England in 1816, by Francis Ronalds. Synchronized clocks with letters on their dials were set up at each end of a wire. Each clock consisted of a rotating disk that had a notch with only one letter visible at any one time. An electric shock, produced by a frictional generator, was sent down a wire. Pith balls became electrically charged and twitched as they repelled each other and a particular letter on the dial would be indicated and transmitted from one clock to the next. Ronalds never secured support to develop his system: The British Admiralty suggested that security needs did not warrant modification of the existing visual telegraph system (Standage 1998, 20).

While financial support for new systems was sometimes lacking, there is evidence that the idea of the electric telegraph, even in its infancy, had captured many imaginations. In 1797, Encyclopaedia Britannica suggested that better communication offered by the telegraph would help facilitate better understanding and greater possibilities for a peaceful society: "The capitals of distant nations may be united by chains of posts, and the settling of those disputes which at present take many months or years may be accomplished in as many hours" (quoted in Standage 1998, 16).

The possibilities of sending messages by using electricity had fascinated a number of scientists. Many had observed that electricity moved almost instantaneously through wires. In 1834 English physicist Charles Wheatstone, who would later, with inventor William Fothergill Cooke, help build one of the first functional telegraph systems, measured the velocity of electricity, coming up with a figure faster than the velocity of light as it was measured then (Rhys-Morus 2000, 459). Electricity was not easy to control in laboratories let alone in "real world" conditions. One of the problems was to be able to produce a steady and reliable supply and to work out ways of producing a predictable signal at one end of the wire that could be reliably received at the other. Developments in the technologies of batteries by scientists such as Alessandro Volta in 1800, and John Frederic Daniel in 1836, helped produce more reliable sources of electricity that would make experiments with the telegraph much easier. Other developments in the science of electricity would also help. Hans Christian Oersted in Denmark in 1820 noted that electricity passing through a wire had an effect on compass needles because it produced a magnetic field, and Joseph Henry (1797–1878) in the United States built and experimented with electromagnets. These scientists discovered that by covering horseshoe-shaped pieces of metal with wires conducting electricity, magnets could be created.

Henry's work was particularly important for the telegraph development. He experimented with different sizes of coils of wires and wrote in one of his scientific papers that his experiments would be helpful in building a telegraph. Henry also helped address one of the problems faced by earlier telegraph inventors of establishing ways a signal could be sent along greater distances of wires without seriously diminishing in strength. His work and advice would prove to be crucial to those who built the first working telegraph system and later the telephone. While Henry never sought to develop his ideas into practical inventions, he would later become quite embittered at the lack of credit his theoretical research would be granted by Morse and other telegraph inventors (Hellman 2004, 39–58).

COOKE AND WHEATSTONE

During the 1820s and 1830s numerous scientists and inventors would apply themselves to building telegraph-like devices. One of the first systems actually put to use was developed by the, often strained, English partnership of Cooke and Wheatstone. Refining ideas of Henry and others, their telegraph relied on a system of magnetized needles that would twist according to changes in current. The original system relied on six wires and five needles through which letters of a message were directly transmitted. This system was later simplified and improved. In 1837 Cooke and Wheatstone secured a patent in England for a needle electric telegraph.

Cooke's father was a friend of Frances Ronalds whose earlier proposals for an electric telegraph had been rejected by the British Admiralty. So he was no doubt aware that he would need more than just government support to get the electric telegraph established. The rapid growth of railways was to provide the answer to his needs. Cooke and Wheatstone demonstrated their telegraph to various railway companies; one of the first lines to be built would be a 13-mile link between Paddington and West Drayton supported by the Great Western Railway. Other telegraph lines along railway lines quickly followed. It was quickly discovered that conveying messages by six wires was unnecessarily complex and, that with the use of code, three wires would do the job almost as well. Operators apparently discovered this by chance when they were faced with problems of wire breakages. Business grew and in 1845 Cooke secured, what Ronalds had not been able to do earlier, a lucrative contract with the British Admiralty. His task was to build an 88-mile electric telegraph line between Portsmouth and London (Standage 1998, 45–47).

The telegraph increasingly captured the public imagination. Newspapers reported on Cooke and Wheatstone's demonstrations of telegraphs and marveled at the speed with which the telegraph allowed public announcements to be made. Within 40 minutes of the birth of Queen Victoria's second son, Alfred, on August 6, 1844, the *Times* carried the story announcing its debt to "the extraordinary power of the Electro-Magnetic Telegraph" (Standage 1998, 50). One of the most sensational telegraph public-interest stories focused on was the role it played in the apprehension on January 3, 1845, of John Tawell who had murdered his mistress in Slough. Tawell sought to evade arrest by traveling to London where he could blend in with the bustling masses. His escape plan came unstuck when witnesses telegraphed ahead to police in London reporting that they had seen Tawell board a London-bound train in Slough. Tawell was arrested as he disembarked in London. The *Times* gloated that without the telegraph his arrest would have been much more difficult, and later, after he was convicted and hanged, they would describe the wires of the telegraph as the "cords that hung John Tawell" (Standage 1998, 51). Popular interest in the telegraph even extended to "mediums" suggesting that the telegraph might provide a means to communicate with the dead. The popularity of "spiritualism" in influential circles of society in Victorian England meant that the idea that science and technology could help provide evidence for spiritualism could in return be used, by various promoters of the telegraph, as a way of encouraging investment in its practical development (Noakes 1999, 425–426).

Anticipating that the telegraph was becoming a financial success, a well-known financier and member of Parliament, John Lewis Ricardo, bought into the telegraph business. In September 1845 he established (with Cooke) the Electrical Telegraph Company. This company would buy out Cooke and Wheatstone's patents and help consolidate the telegraph as a "day-to-day" part of the life in Victorian England. In 1869 the company would be absorbed by the post office, marking the beginning of a long era of government control of the telegraph in Great Britain (Standage 1998, 56, 161).

SAMUEL MORSE

At around the same time as Cooke and Wheatstone, Samuel Morse from New York would also develop an electric telegraph system. Morse was to become the inventor who would draw the greatest fame from the development of the electric telegraph and the one best remembered by history. The extent of his contributions would, nevertheless, be subject to dispute in his own time. One of Morse's contemporaries and one-time supporter, congressman Francis O. J. Smith, was later to demean Morse's individual contribution: "While Henry was strictly the father, Gale was no less truly the midwife, at the birth of the American Electromagnetic Telegraph. Professor Morse, in fact only acted the part of the errand boy, who called in the midwife's service, to save the life of the unborn child; and even after its birth, it was too feeble and slow of motion, too deformed in limbs and speech, to be of value, without the nursing and ingenious new mechanical appliances of Mr. Vail, or some equivalent artisan, and it is sure that without them it could never have grown into manhood, or have been utilised for business purposes" (quoted in Hellman 2004, 55). Smith correctly notes that the telegraph was an invention that relied on the ideas and work of a number of individuals, but it is a view that underplays Morse's creativity, imagination, and capacity to link people and ideas together. When the issue of Morse's originality came before the Supreme Court in 1853, they found in his favor, upholding his patent (Standage 1998, 171–172).

Samuel F. B. Morse was born in Charlestown, Massachusetts, in 1791. He attended Yale University, where he was an ordinary student burdened by debts acquired through excessive partying and drinking (Schwartz-Cowan 1997, 124). Morse turned out to be a better artist than a scholar and set out on a career as a painter, achieving some success in portraiture. A sad set of events that may have stimulated Morse's interest in the telegraph surround the circumstances of the death of his wife, Lucretia. She passed away suddenly on February 7, 1825, at their residence in New Haven, Conneticut. At the time, Morse was in Washington, some 4 days of travel away. He only received notice of her death on February 11. In the meanwhile, on February 10, in a letter which must have crossed paths with the tragic news from New Haven, he had written to his wife "I long to hear from you." Despite his best efforts, Morse did not manage to get back to New Haven in time for Lucretia's funeral (Standage 1998, 26). It almost seemed that space and time had conspired to intensify his loss.

Morse's painting and teaching afforded him a living, but not the wealth and fame that he sought, nor the opportunity to paint the way he desired. Morse had always been interested in tinkering and inventing and increasingly turned his efforts in this direction. The basic elements of Morse's electric telegraph emerged in 1832, when he was returning to the United States from Europe aboard the ship *Sully*. These journeys could take many weeks, so Morse had plenty of opportunity to read and discuss with his fellow travelers the new exciting theories of electricity. Morse filled books with sketches and notes about how an electric telegraph might be built. His enthusiasm was struck a blow when shortly after his return to the United States he became aware that a number of other inventors and scientists had already begun to develop telegraphs. So he temporarily shelved his plans. Morse continued his artistic career with some success securing, in 1835, an appointment as Professor of the Literature of the Arts of Design at New York University (Lubar 1993, 76).

In 1837, Morse regathered his thoughts and finally decided to try to build his own electric telegraph. Lacking specialized knowledge of the theories of electricity, he sought out Leonard Gale, Professor of Geology and Mineralogy at New York University, the same university where Morse taught art. Gale became a partner with Morse providing scientific advice in return for a share in profits and patents. Gale also introduced Morse to Joseph Henry. As was noted above, Henry was a leading scientist working on the new sciences of electricity and magnetism. Henry, who was teaching at Princeton University, adopted the attitude that it was not for scientists to receive patents. He supported Morse asking for nothing in return. It would prove to be a relationship that would later be tinged in bitterness. Henry helped Morse by providing more specialized theoretical knowledge about electricity and magnetism. Henry is often attributed with the statement that he found Morse to "have very little knowledge of the general principles of electricity, magnetism, or electro-magnetism" (Lubar 1993, 77).

Morse's electric telegraph relied on a number of simple key ideas: electricity passing through a coil of wire produces a magnetic field and that the presence of the field can be detected by a piece of metal or needle. If electricity passing through a wire is interrupted then so will the magnetic field. Using a switch to disrupt the flow of electricity produced by a battery in one place can be detected at the other end of a wire and be used to transmit information. The details involved in getting these simple ideas to actually be useful involved solving numerous theoretically basic, but practically demanding, problems. These included working out ways of improving the strength of existing batteries to guarantee a more reliable source of electricity; ways of passing electricity through significant distances of wire without energy petering out; building switching devices that would interrupt the flow of electricity and magnetic fields in a controllable way; developing detection devices that could register the interruptions to a magnetic field in a dependable way; and methods of encoding messages so they could be sent using a basic signal.

In his earliest attempts to build a system Morse relied on readily available and familiar materials from his art workshop and his first code relied on assigning numbers to words. The sender and receiver both had a codebook. Sending a message relied on the sender tapping out numbers with larger spaces between separate numbers and smaller spaces between the individual digits of numbers. This system was significantly improved when he enrolled the assistance of Alfred Vail. Vail had been a student at New York University, and from experience in working in his father's ironworks

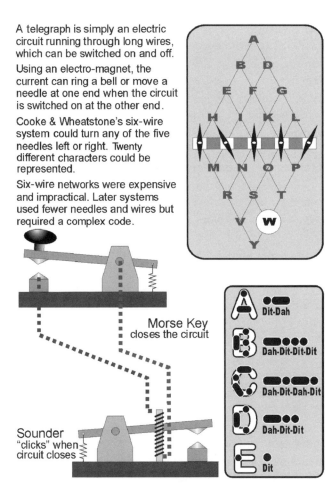

A telegraph is simply an electric circuit running through long wires, which can be switched on and off.

Using an electro-magnet, the current can ring a bell or move a needle at one end when the circuit is switched on at the other end.

Cooke & Wheatstone's six-wire system could turn any of the five needles left or right. Twenty different characters could be represented.

Six-wire networks were expensive and impractical. Later systems used fewer needles and wires but required a complex code.

Morse Key
closes the circuit

Sounder
"clicks" when
circuit closes

A ●■ Dit-Dah

B ■●●● Dah-Dit-Dit-Dit

C ■●■● Dah-Dit-Dah-Dit

D ■●● Dah-Dit-Dit

E ● Dit

Morse and Vail developed a single-wire code of short and long impulses for letters and numerals.

This code could be heard by using a clicking sounder or recorded for later translation by embossing marks onto a strip of paper.

Early electric telegraphs, Cooke and Wheatstone and Morse. Courtesy Robert B.K. Brown, 2006.

was more mechanically proficient than Morse. Vail agreed to assist Morse in improving the construction of the telegraph in return for a quarter-share of future profits. With Vail's assistance a new code was developed that was much quicker than looking up a word for every number. Morse's original ideas for code had been too attached to his studies of the Chappe system. The new code relied on two main types of signals, combinations of "dots and dashes" that would correlate with letters from the alphabet, rather than numbers. More common letters had shorter codes.

The code proved to be relatively easy to learn and didn't require lengthy codebooks, and allowed messages to be transmitted reasonably quickly, up to 30 words per minute. In 1840, Morse obtained a U.S. patent for his electric telegraph. Two years earlier he had traveled to Europe and been unsuccessful as the alternative electric telegraph of Cooke and Wheatstone had already been patented in England. Morse was disappointed that the U.S. government didn't buy his patent, but he did achieve some success in gaining government support mainly through the lobbying of the congressman Frances O. J. Smith who was chairman of the Committee of Commerce. Smith had also secretly become a quarter partner in the U.S. telegraph patent with Morse, Gail, and Vail. In a report to Congress, Smith used utopian images to celebrate the political possibilities of the telegraph: "The influence of this invention over the political, commercial, and social relations of the people of this widely extended country . . . will . . . of itself amount to a revolution unsurpassed in moral grandeur by any discovery that has been made in the arts and sciences. . . . Space will be to all practical purposes of information, completely annihilated between the States of the Union, as also between the individual citizens thereof" (quoted in Lubar 1993, 80–81).

With Smith's support and continued lobbying Morse gained financial backing, to the tune of $30,000, to build a telegraph between Washington, D.C., and Baltimore. Building an operational line offered numerous challenges. Borrowing from Cooke and Wheatstone, Morse and Vail found that it was easier to attach telegraph wires to poles rather than pass them underground as they had originally planned. They also developed a simple spring-loaded keying device for sending messages and improved ways of recording messages. In 1844 the first official message was sent from Baltimore to Washington, setting the tone for the occasion. Morse's mundane dots and dashes spelt out the portentous words "What hath God wrought." Further successful messages were sent from Baltimore to Washington relating news of presidential ticket selections. These were received with enthusiasm, but the market for the telegraph was yet to develop, the commercial value of certain forms of information were not yet clear, and the government failed to continue to subsidize the line. In 1845 the partners in the patent began to sell their shares. This led to the emergence of a number of private telegraph companies.

Morse's vision had been for the government to play a stronger role in regulating the telegraph and avoid the traps of monopoly private ownership. The idea that the government should be involved in maximizing the broader community benefits of the telegraph provided a different model for the role

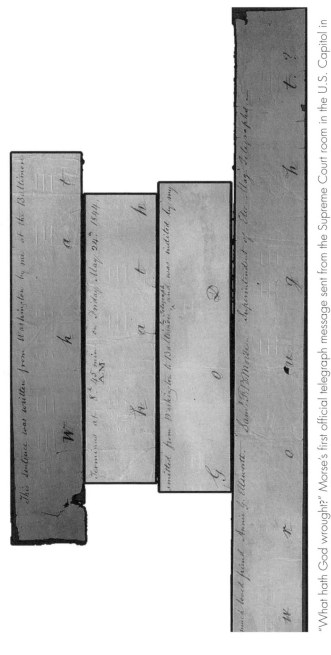

"What hath God wrought?" Morse's first official telegraph message sent from the Supreme Court room in the U.S. Capitol in Washington to his assistant Alfred Vail in Baltimore, on May 24, 1844. Courtesy of the Library of Congress.

of government to that of the Chappe telegraph systems. Rather than be used as a tool of surveillance and political control, Morse suggested that the telegraph should be encouraged to take on a distinctive American form, an "American Telegraph" reflecting a democratic American political vision (John 1998, 196–197). While the U.S. government would not provide direct ownership it still played an important role in subsidizing the expansion of the telegraph as a way of encouraging nation building. In 1860, for instance, Congress passed a bill "for facilitating communication between the states of the Pacific and the Atlantic by means of the electric telegraph" (Flichy 1995, 42).

Morse's work had a huge influence across the world. While Cooke and Wheatstone had developed the first working telegraph system and held the British patents for the telegraph, the simplicity and effectiveness of Morse's code would mean that it gradually became widely used in their system anyway, and in Europe, Morse's system became the standard from 1851. Despite these successes, it took Morse a number of years to be given recognition in his home country and while he was bestowed numerous ceremonial honors across Europe, because he had been unsuccessful in taking out European patents (with the exception of France), he initially received very few financial rewards from them. With the final success of the transatlantic cable in the 1860s, and lobbying by the emergent telegraph industry in the early 1870s, Morse, by now an old man, would finally receive official acknowledgment in the United States. On June 10, 1871, Morse was hailed as the father of the telegraph with the unveiling of a bronze statue in Central Park, New York (Standage 1998, 170–176).

THE GROWTH OF THE ELECTRIC TELEGRAPH

As noted earlier, the electric telegraph would spread in Great Britain mainly through its attachment to the boom of the railways. By 1848, telegraph wires ran alongside approximately 50 percent of railways, and by 1850, Britain could boast 2,215 miles of wires. The telegraph was also steadily adopted by other nations: in Prussia, there was 1,493 miles of wire; in Austria 1,053 miles; in Canada 983 miles; with other nations following quickly. France offered a partial exception, possibly as throwback to its success with the Chappe telegraph having only 750 miles in 1852 (Standage 1998, 60–61). While U.S. markets for the use of the telegraph were initially slow, they quickly grew until the use of the telegraph was unparalleled by any other nation. By 1846, 1,200 miles of telegraph lines extended across the United

On June 10, 1871, Morse, by now an old man, was finally officially honored as the father of the telegraph. "Morse self-portrait." Courtesy of the Library of Congress.

States; by 1850, 12,000 miles; by 1852, 22,000 miles, and by 1854, there were over 30,000 miles of lines across the country (Flichy 1995, 42).

One of the early stimuli for the extension of the telegraph in the United States was to relay news quickly, particularly news of war and political conflict: newspapers emerged as an important early market. These customers were quickly augmented by other businesses. Regular telegraph messages could report on the progress of trains and possible sources of delay along their line. Businesses dealing with perishable products could secure benefits from timely information about delays in delivery. The development witnessed the amalgamations of a number of smaller companies in 1866 to create Western Union which largely specialized in short business messaging. It became the first U.S. firm to span the whole continent.

THE EIGHTH WONDER OF THE WORLD: THE TRANSATLANTIC CABLE

As knowledge and experience helped telegraph lines to spread, one of the most ambitious technological projects of the nineteenth century was undertaken, the transatlantic telegraph cable christened as the "Eighth Wonder of the World." In 1851, the first undersea cable was constructed between England and France. Experiments involving the sending of messages through water, and by the use of different types of submerged cables, had been a matter of scientific curiosity for some time. Morse, for instance, back in the early 1840s, had experimented with sending current through a wire across the New York Harbor; and Wheatstone had done similar experiments around the same time across the Thames in London (Standage 1998, 67). By 1852, the radical proposal of providing a cable to join the two great Anglophone nations, England and America, was mooted. Utopian visionaries suggested that a lack of communication was one of the key sources of human conflict. For them, the extension of the telegraph was both a resource to unify the United States and also an opportunity to help bind together the United States and Britain. There were also more mundane visions of profits that might be made through providing a means for transatlantic businesses and investors to more efficiently communicate and conduct business, as well as the anticipated demand for more timely reporting of news between Europe and the United States.

One of the key instigators of the transatlantic cable was the New York businessman Cyrus Field. Field was born in Stockbridge, Massachusetts, on November 30, 1819. He made fortunes from paper manufacturing and, by the late 1830s, he was one of New York's wealthiest businessmen; his energy and enthusiasm would be critical for the development of the telegraph. During the early 1850s, there had been speculation that the best route for a transatlantic submarine telegraph link would be from Newfoundland to Ireland. These ideas were given support when surveys done in 1853 indicated there was an underwater plateau in the Atlantic Ocean roughly corresponding with the proposed route.

A telegraph company manager, Frederick Gisborne, began building a line but his early efforts met with little success. So he set out to gain better support and funding. He managed to gain the attention of Field who, after seeking advice from Morse and others, decided to throw himself into developing the project. To begin, Field had to raise finance to extend existing telegraph lines that ran between New York and Nova Scotia to St. John's, Newfoundland. After securing finance, and successfully pursuing this project, the next more challenging task of crossing the Atlantic lay ahead. In

1857, with some finance from the British government and wealthy private investors, and the begrudging support from the U.S. Congress (some anti-British senators believed the project was unpatriotic), Field created the Transatlantic Telegraph Company. A treaty between the United States and Britain was entered into and the project began in earnest.

The cable would be made from copper wire insulated by a tough form of latex known as gutta-percha (early golf balls were made from the same substance) and protected with heavy iron wire. The cable would weigh too much for any single ship to carry. So the company drew upon the services of the largest ships of the United States and British navies: the USS *Niagra* and the HMS *Agamemnon*. The first attempt at laying a cable started out from Valentia Island in Ireland in August 1857, but was a failure marred by repeated breakages of the cable. In June 1858 another attempt was made. The *Niagra* and *Agamemnon* were to meet in the middle of the Atlantic, splice the two halves of the cable together, and then head in the opposite directions to Newfoundland and Valentia Bay. Yet again these attempts were plagued with difficulties; bad weather (and even whales) and further cable breakages occurred. Yet again, the attempt was abandoned and the ships returned to port to take on new provisions. Finally, setting out in July 1858 the same plan was revisited but this time with "success." On August 5 the cable was "landed."

On August 17, 1858, the first official transatlantic telegraph message was transmitted from Queen Victoria, in England, to President James Buchanan, in the United States. Capturing the effusive optimistic tones that would follow in the popular press, the president declared that the telegraph "is a triumph more glorious, because far more useful to mankind, than was ever won by conqueror on the field of battle" (quoted in Standage 1998, 79). The actual transmission was far from straightforward; the telegraph was highly unreliable and the message took 16.5 hours to reach Washington and another 10 hours for the return message to reach London. The quality of signal from the telegraph rapidly deteriorated over the next month and, finally, failed altogether when, in an attempt to improve the signal, a higher voltage was applied to the line. During its first ill-fated month of operation the telegraph had been the source of public amazement and fanfare. For instance, over 15,000 New Yorker's (for that time a massive attendance) turned out on September 1, 1858, to celebrate in a public daytime parade down Broadway. The parade was followed by an evening torchlight procession and a fireworks display (which caused a fire in City Hall). Songs and poetry were composed to commemorate the occasion, souvenir pendants made from pieces of cable, and even a perfume dedicated to Cyrus Field "distilled from ocean spray and fragrant flowers" was sold (quoted in Kennedy 2005). Enthusiasm was not restricted to New York, in London the *Times* reported: "Since the

FESTIVAL SONG

AT THE CELEBRATION OF THE LAYING OF THE

ATLANTIC TELEGRAPH.

New York, on the 1st day of September 1858.

Dedicated to the „Atlantic Telegraph Company" by

WILLIAM SPITZNASSKI.

TRANSLATED FROM THE GERMAN.

Hail man's great intellectual power! —
This blissful gift from God on high
Has brought to earth, a blessed dower,
Electric lightning from the sky.

The heav'nly spark has FRANKLIN captured,
MORSE sent it speaking through the land,
And now has FIELD two worlds enraptured,
By spreading it from strand to strand.

World's free Lord, man, it is who measures
Earth's caverned, rocky depths below;
He brings to light earth's secret treasures,
To banish misery and woe.

Metallic chord his skill is drawing,
To lay it on deep ocean's sand,
To send beneath the tempest's howling
The word of peace from land to land.

Peace be on earth to ev'ry nation!
We sing to human mind's great praise,
One harmony is all creation,
One family the human race.

This day we hear the cannon's sounding,
We hear the bells so charming ring,
And all our hearts, with joy abounding,
In Union with all nations sing:

"One heav'nly spark unites for ever
The earth and ev'ry human breast;
All men are brothers wheresoever
Our eyes upon the earthball rest!"

Parades, fireworks, and public celebrations were held on September 1, 1858, to herald the opening of the transatlantic cable. Even special songs were composed for the occasion. "Festival song at the celebration of the laying of the Atlantic Telegraph." Courtesy of the Library of Congress.

discovery of Columbus, nothing has been done in any degree comparable to the vast enlargement which has thus been given to the sphere of human activity" and "the Atlantic is dried up, and we become in reality as well as in wish one country" (quoted in Standage 1998, 80–81).

When it became known that the telegraph was no longer working there was, as could be expected, a strong backlash. There was even speculation that the whole thing had been an elaborate scam. In response, a joint American and British committee of inquiry, including among its members Wheatstone, was established to investigate the telegraph's failure. One of the most important expert witnesses the committee would call upon was William Thomson (later to be knighted and known as Lord Kelvin). Thomson was Professor of Natural Philosophy at Glasgow University and would come to be one of the most respected scientists of his time becoming well known for his contribution to developing the theory of thermodynamics. Kelvin recommended a much larger conducting core for the cable, and that it should be made more buoyant so it would be less prone to breaking under its own weight. He also developed a much more sensitive device for reading the faint signals that the submarine telegraph transmitted: the mirror galvanometer. By using this device, and a better conducting core, lower voltages could be used, which reduced the problems of the earlier cable where the use of higher voltages to try to improve the strength of signal had damaged the cables insulation (Standage 1998, 83–84).

The new improved cable was almost twice the weight of the old. So the largest ship in the world, *The Great Eastern*, which had proved until this time to be something of a "white elephant" (it was simply too big to be particularly useful), was prepared for the task. On June 24, 1865, she set out on what was to be yet another failed attempt to establish the line; two-thirds of the way across the Atlantic the cable snapped. Just over a year later, on Friday, July 13, 1866, another attempt was made, but this time it would be an unqualified success. *The Great Eastern* not only smoothly laid the new telegraph line but also a month later recovered and repaired the lost cable from 1865. There were now two working transatlantic cables. The enthusiasm of 1858 was revisited. The pioneers of the telegraph, Cooke and Wheatstone, and Ronalds were bestowed with various honors; Thomson was knighted; Field received a specially minted gold medal from the Congress; and lavish banquets were held for Morse in New York. One of the British ambassador's toasts to Morse waxed lyrical: "The telegraph wire, the nerve of international life, transmitting knowledge of events, removing causes of misunderstanding, and promoting peace and harmony throughout the world" (quoted in Standage 1998, 87). It is worth noting that these triumphs of the telegraph were set against the intense historical

Despite the rhetoric that the telegraph was a tool for peace, the U.S. Civil War showed that the telegraph could also be used to coordinate even more intense military campaigns. "Field Telegraph Station: U.S. Civil War 1861." Courtesy of the Library of Congress.

backdrop of the U.S. Civil War that raged from 1861 to 1864. Despite the rhetoric that the telegraph was a tool for peace, the Civil War showed the telegraph's other potentials, better communication could also be used to coordinate ever more intense and bloodthirsty military campaigns.

ANNIHILATING SPACE AND TIME, THE TELEGRAPH AND BROADER SOCIAL AND ECONOMIC CHANGE

The successful spread of the telegraph has often been linked to shifts in perceptions of time and space which can, in turn, be linked to things like new

The telegraph would continue to be an important form of telecommunication well into the early twentieth century. "Western Union Telegraph Delivery Boys," 1916. Courtesy of the Library Congress.

forms for the coordination of institutions of government and business, standardization of time and new genres for reporting news. Large institutions had been long aware of the importance of transport and communication in coordinating their activities. Steady improvements in roads, canals, shipping, ports, postal services, and railways, all provided ways of enhancing their powers of coordination and control. The telegraph fitted in with this framework well, but, in particular, it influenced the methods of coordination by offering the possibility of "shrinking" time and space. Telegraphic communication was far less constrained by geography than traditional forms of communication. Communication with distant colonies and dominions was now almost instantaneous, "global" trade could be better organized, troops more quickly dispatched, and telegraph lines provided symbolic reminders to scattered imperial subjects of the sovereign's omnipresence (Bektas 2000, 669). Large businesses also exploited these potentials being better able to directly control their activities at a distance from their head offices. Already large cities, like New York and Chicago, could grow further as centers of business calculation and coordination (Nye 1997, 1072–1075). The importance of coordinating business from these centers meant that, increasingly, information was viewed as a commodity in itself. For example, being the first to know about prices on the market for stocks could enhance profits or minimize risks. As messages could move faster than transportation, traders also needed to consider prices in terms of future market conditions. These abstract forms of trading in turn also encouraged standardization of products and time. To coordinate railways and trade, standard time zones were also encouraged: contemporary with the spread of the telegraph, on November 18, 1883, a grid of hourly time zones was imposed on the United States (Carey 1989, 316–317).

The capacity for the telegraph to shrink perceptions of time and space also contributed to changes in the genre of news reporting. Distant events were reported "more or less" as they happened and disseminated to wider audiences. The telegraph's relatively limited capacity to carry large amounts of information also required an economic use of language. Reporters became aware that they were reporting to more geographically dispersed audiences for whom the imagined world was closer and, in a sense, smaller. This encouraged them to use language and reporting styles that were less local and idiosyncratic than in the past. Stories needed to be written with an imagined national community in mind (Moore 1989, 31–34).

The embrace of the telegraph as a tool to help coordinate commerce and disperse news, particularly in the United States, had made it a lucrative business. In 1870, William Orton, president of Western Union, who more or less held a monopoly on the U.S. telegraph business, suggested to a congressional committee: "the fact is the telegraph lives upon commerce . . . it is the nervous system of the commercial system. If you will sit down with me at my office for twenty minutes, I will show you what the condition of business is at any given time in any locality in the United States" (quoted in Standage 1998, 160). It is not surprising that by 1880 a lucrative U.S. telegraphic market could boast 291,000 lines (Lubar 1993, 91).

2

The Invention of the Telephone, 1876

◆

The word telephone comes from the Greek word "tele" meaning afar and from "phone" meaning voice. Alexander Graham Bell is normally recognized by history as the inventor of the telephone. It is nevertheless important to remember that the process of the invention of the telephone took a number of years and involved a number of other inventors whose contributions came close to matching Bell's. These questions in fact arose during the early years of the telephone's development and featured in a later tangled web of 18 years of litigation that involved Bell's patents being tested in 600 separate cases (Bruce 1973, 271). Bell's claims to priority to the invention may have been slightly tarnished but ultimately survived legal challenge. All courts found in favor of Bell, but one case that went to the U.S. Supreme Court, brought by the Attorney General accusing the Patent Office of impropriety and Bell of obtaining his patent by fraud, was never formally settled. It dragged on from 1887 to 1896 and was finally dropped on the basis of a lack of funding and legal technicalities (Bargellini 1993, 417; Bruce 1973, 275–277).

These cases helped bring to light numerous claims that probably otherwise would have been relegated to the dust of history, and it is no doubt true that many of the claims were not legally sustained because they were opportunistic and lacked substance (Bruce 1973, 271). Not all were suspect nevertheless, and the claims of Elisha Gray stand out as worthy of serious analysis. It is also important to acknowledge that Thomas Edison's slightly later contributions were crucial to the development of a practical working telephone. There were also a number of other less widely known inventors, contemporaries of Bell, whose contributions are not easy to verify, nor totally reject. For instance, the Italian American inventor Antonio Meucci's (1808–1889) work has often been dismissed out of hand (Coe 1995, 39–46; Bruce 1973, 271–272) but some historical reconstructions suggest that his work displayed some ingenuity and that his claims may have been treated unfairly (Bargellini 1993, 419–420). Because of his language problems, poor finances, and ill health, Meucci could not afford to maintain or pursue patents on his telephone ideas. These problems would resurface in legal proceedings against Bell where Meucci's briefs were poorly prepared and he required an interpreter to provide testimony. Bell's counsel went as far as challenging Meucci's credibility on the basis of his ethnicity, describing him as "an imposter, a man who had little control over his sense of truth, a Latin—not an Anglo-Saxon" (quoted in Bargellini 1993, 418).

ORIGINS OF THE TELEPHONE

Broadly speaking, the idea of the telephone has multiple origins. In the early days of Western Science, Frances Bacon in his book *New Utopia* (1627) described a telephone-like device involving a long speaking tube. Slightly later, in 1667, Robert Hooke, perhaps best known for his debates with the great Isaac Newton, performed experiments involving the transmission of sound along a taut string. Devices using cups attached to either ends of a string also became common in the nineteenth century and were known as "lover's telegraphs." Of more immediate significance, in terms of using electricity to transmit sounds, was the work in 1837, of American physicist William Charles Page. Page explored the possibilities of producing sounds by rapidly magnetizing and demagnetizing metal rods; the sounds that they emitted displayed a relationship to the rate at which the rod was magnetized/demagnetized. Page's work would provide an inspiration to Reiss, Edison, Bell, and other later telephone inventors. In the 1850s in France, Belgian-born inventor Charles Bourseul presented a report to the *Academie Des Sciences*, which discussed the possibility of transmitting sound vibrations via electricity. His 1854 discussion described extending the telegraph to transmit speech. This would be done by using a flexible disk-like device that would break or make a connection with a battery and in doing so vibrate with speech-like sound vibrations. Bourseul's work also became widely known although he never actually built a working model of his proposed device (Flichy 1995, 82–84).

The first demonstrations of a "telephone-like" device were by Johan Philip Reis at the Physical Society of Frankfurt, Germany, in 1861. He had built his model by drawing analogies with the physiological structures of the human ear, something that Bell would also later do. Reiss constructed a Bourseul-type of device, although whether or not he was aware of Boursel's work has been open to debate (Flichy 1995, 83). The transmitter was composed of a vibrating diaphragm with contacts made of platinum that made or broke a circuit with a battery. The receiver was composed of a coil of wire wrapped around something like a knitting needle fastened to a sounding board.

Most of these early devices relied on variations of the Page effect described above (Meyer 1995, 4–5). By relying on the Page effect, and thinking as telegraph inventors, the common assumption was that speech would be able to be to be conveyed by on and off currents. Because speech has the quality of a fluctuating continuous wave and not a collection of Morse code-like pulses, these systems would always face limitations transmitting speech. It would not be until the recent digital era that the natural patterns of speech can be sampled and computer encoded at a sufficient rate to be transmitted by pulses. There is some speculation that Reiss's apparatus may have occasionally come close to transmitting speech "accidentally" when it was conveying a signal that was so weak, or if it was "malfunctioning" (such as corrosion sticking metal contacts together), that the contact connecting the circuit to the transmitter was constant. Reiss's work was well known to Gray, Bell, and Edison. Joseph Henry brought back to the United States, from Europe, a copy of a Reiss telephone-type device. Bell saw this device in a visit to Henry at the Smithsonian Institute in March 1875 (Flichy 1995, 83).

German scientist Hermann von Helmholtz (1821–1894) also explored the possibility of using electricity to send complex signals down a wire. Helmholtz had managed to transmit vowel-type of sounds using a combination of resonators and electrical tuning forks. While most telegraph inventors lacked a detailed scientific understanding of Helmholtz's work (assisted by its then limited availability in English translation), it provided part of the informal scientific backdrop that inventors drew inspiration from, along with the work of scientists such as Joseph Henry (Bruce 1973, 50–51).

The emerging electrical sciences were one of the stimulants for the invention of the telephone, but it is important not to assume there was a simple, or automatic link, between these new forms of knowledge and actual inventions. Inventors would often borrow bits of the science that were useful to them and ignore, or be unaware of, others. They would also reformulate scientific principles in practical terms and produce devices and effects that went beyond the scientific understandings of the time (Hoddeson 1981, 516–519).

Different inventors drew on science in different ways. For instance, Gray had some formal scientific training directly relevant to electrical invention. His attitude to invention was professional and pragmatic, and, as a well-respected highly successful inventor, he shaped his research strongly according to its anticipated commercial viability and toward maintaining his high standing among his peers. Bell's professional training was in the field of what would be described today as elocution and speech therapy and his scientific knowledge of electricity was uneven and, in some areas, quite limited. Bell was nevertheless quite open in seeking scientific opinions to help his projects. While, like Gray, Bell was keen to secure financial rewards for his research, he seemed to find it more difficult than Gray to suppress what would prove to be a lifelong diverse sense of curiosity. Bell's relative lack of specialist knowledge and lesser standing as an electrical inventor also meant he was able to take his research into the "speaking telegraph" more seriously than Gray who largely treated the topic as a novelty (Hounshell 1975, 160–162). Many years after the invention of the telephone, in a 1913 address to the 3rd Annual Convention of Telephone Pioneers in Chicago, Bell's longtime assistant, Thomas Watson, would note ironically, "[if] Bell had known anything about electricity, he would have never invented the telephone" (quoted in Bargellini 1993, 410).

The other more immediate source of inspiration for the invention of the telephone were attempts by numerous inventors, of whom Bell was but one, to reap profits from patenting improvements to the telegraph that had become a booming industry. Enos Barton, who would with Gray's help, found Western Electric, later a giant in developing telephone technology, reminisced on the sense of opportunity and anticipation of the time "[f]ortunes had been made in developing the telegraph system, and it was the general expectation that there were other fortunes awaiting the development of new inventions. The electrical inventor could easily get the ear of the capitalist, and the capitalist even sought out the inventor. The decade from 1870 to 1880 saw the beginning of many great things in electricity" (quoted in Young 1983, 1).

One of the key problem areas of technological interest for the telegraph industry was to provide solutions to the problem of sending more than one message down a telegraph line at the same time. The copper wiring required for telegraph lines was becoming increasingly expensive, and multiple lines were unsightly and even dangerous, coming down in storms. Increasing the capacity of existing wires would save costs and help stop congestion. William Orton, the boss of Western Union, reputedly had on offer $1 million to the inventor who could develop a system for sending multiple telegraphs (Hounshell 1975, 144).

Western Union had become the most powerful telegraph company and exerted a monopoly on what had become a lucrative industry. It took a keen interest in tracking innovations that would improve telegraphy, not only to promote but also delay them, depending on its business interests. William Orton, who took over running Western Union in the 1870s, purchased the rights to the patent for a system, developed by Joseph B. Stearns that could send two messages simultaneously down the one line. In 1872, Orton described the Stearns Duplex "as the single most important invention in telegraphy since Morse" (quoted in John 1998, 201). Orton went on to hire Thomas Edison (1847–1931) to work on the problem of multiple messaging. Edison went on to develop a "quadraplex" (4-message system) and, later, contributed to significantly improving early telephone design. One of the main approaches for attempting to solve the problems of multiple messaging was the development of so-called harmonic telegraphs. Both Bell and Gray would place considerable energy into trying to develop these types of devices.

ELISHA GRAY, THE MAN MOST LIKELY TO INVENT THE TELEPHONE

Elisha Gray was born on a farm in Barnesville, Ohio, in 1835. His father's death meant he was forced to leave school, for work, at 12 years of age. It was not until in his twenties that he would overcome hardship and return to schooling, ultimately finding his way to Oberlin College, where he studied electrical sciences with Professor Charles Churchill. Throughout his life, Gray experienced significant periods of poor health. His sickliness appeared to allow him to focus his mind rather than offer a source of distraction. By the age of 32, he embarked on a career as a professional telegraph inventor experiencing success with his first patent being granted in 1867. From the time of his first patent, Gray was noticed by the telegraph industry and had a run of successful inventions, such as improved telegraph printers. Gray invested his profits in setting up a telegraph instrument manufacturing company with another telegraph inventor, Enos Barton. In 1870, their company, Gray and Barton, would, with the support and investment of William Orton, become Western Electric Manufacturing Company. Gray was for a time its superintendent and a board member (Hounshell 1975, 137).

Like the mythical accounts from the history of science of Archimedes, the humble bathtub would provide an important inspiration for Gray's telephone work. In early 1874, Gray's nephew was playing with some of his uncle's electrical equipment in the bathroom, deliberately administering himself electrical shocks. He did this by connecting the zinc wire from an induction coil, which turns the direct current from a battery into continuously alternating current, to the zinc lining of the bathtub. Another wire was held in one hand and the free hand was rubbed along the bath so as to achieve the desired electric shock. While this was happening, parts of the induction coil vibrated producing an identifiable pitched sound. Gray noticed that where his nephew's hand rubbed along the bath the same pitched sound was being reproduced. Moving his location, the coil, and nephew, different pitches were produced from the coil but continued to match the pitches produced by the hand rubbing the bath. Gray drew considerable inspiration from this. It suggested to him that it was possible for a known pitch, or frequency, to be electrically transmitted and received (Hounshell 1975, 138–142).

Soon after the "bathtub experiments" Gray resigned from his directorship at Western Electric and began devoting his full-time energies to building electrical transmitting and receiving devices. He built devices to transmit single and two tones, and for receivers, built ingenious combinations of vibrating metal plates attached to violins and vibrating diaphragms made from shoe polish cans that substituted for the bath. Gray also began developing ways of electrically transmitting musical tones with single-toned generators tuned to different notes of the musical scale. In 1874, he demonstrated this device to an audience from the telegraph industry and later went on to build a one octave (eight notes) musical transmitter built up of eight single-toned transmitters. These were triggered by a keyboard, and sent musical tones that were received by a washbasin mounted near the poles of an electromagnet.

These early attempts to send multiple signals encountered the problem of capacitance: the signals became scrambled as they passed along the telegraph wire. This necessitated Gray working on the development of receiving devices that could unscramble them. By 1875 Gray had addressed many of these challenges and filed a number of patent applications for harmonic telegraph devices. Around the same time, Gray also became aware that Bell was "hot on his heels" trying to invent a workable harmonic telegraph and was also interested in the telegraphic transmission of speech (Hounshell 1975, 148–152).

On February 14, 1876, Gray would file a patent caveat: this was a notice of an inventor's idea that would shortly be built into a practical device that could be patented. Gray's caveat was for a so-called "talking telegraph," drawing inspiration from the "lovers telegraph" (a popular novelty device where two tin cans were linked by string). Gray had thought about ways of electrically, rather than just mechanically, transmitting voice along a wire. Gray's caveat described a voice chamber with a diaphragm at its base. This diaphragm would respond to the vibrations made by the voice. The diaphragm would have a wire connected to it that would be dipped in a solution connected to an electric circuit. In response to the vibrations of the voice the wire would dip either deeper or more shallowly in the solution and in turn, increase or, decrease the resistance of the electrical circuit (Hounshell 1975, 152–154).

In an extraordinary "coincidence" only hours before Gray's caveat had been lodged a patent application had been lodged on behalf of Bell, also for a speaking telegraph device. In legal circumstances that would become more confused when contested later. The Patent Office decided to award Bell's patent and not issue an interference against Gray (this would have given Gray the option to present his case for priority). Gray could have nevertheless still "pushed the issue," and made sure an "interference" was acknowledged if he had been willing to straightaway apply for a patent.

Gray's lawyer advised: "Gray's 'Talking Telegraph Caveat' interferes with an application of Bell's, but as Gray's caveat was filed the same day as Bell's application but later in the day, the Commissioner holds that he is not entitled to an interference, and Bell's application has been ordered to issue. . . . We could still have an interference by Gray's coming down tomorrow and promptly filing an application, for a patent. If you want this done, telegraph me in the morning, on receipt of this, and I will have the papers ready in time to stop the issue of Bell's patent; but my judgment is against it . . ." (quoted in Hounshell 1975, 154).

Gray agreed with his legal advice, and later in 1876, would a number of times trivialize the significance of Bell's work, "Bell has talked so *much* and done so little *practically*. I am working on an Octoplex between Philadelphia and New York—four [messages] each way simultaneously—eight [messages] at once. I should like to see Bell do that with his apparatus" . . . "[the] talking

Alexander Graham Bell maintained an intense curiosity in science and technology throughout his life. Portrait of A.G. Bell. Courtesy of the Library of Congress.

telegraph is a beautiful thing in a scientific point of view. . . . But if you look at it in a business light it is of no importance. We can do more. . . . More with a wire now, than with that method" (quoted in Hounshell 1975, 157). After a deceptively slow commercial start when Bell's telephone began to suggest that it would be a commercial success, Gray would desperately try to recover his lost opportunities in bitter legal proceedings with Bell.

ALEXANDER GRAHAM BELL, THE SPEAKING TELEGRAPH (TELEPHONE) IS BORN

Alexander Graham Bell was born in 1847 in Edinburgh, Scotland, into an educated middle-class family. His family background would exert an influence on his work throughout his life. His father and grandfather had both been experts in elocution and the study of speech. Bell's grandfather (Alexander Bell) had operated a school for elocution and Bell's father (Alexander Melville Bell) invented a teaching system known as visible speech. This system used thirty-four written symbols to capture verbal sounds: the symbols displayed placement of the tongue, throat, and lips during speech. It was designed to help in the teaching of foreign languages and became a tool for teaching the deaf. A. M. Bell's textbooks were well known across Britain and the United States—so much that he received an acknowledgment in the preface to George Bernard Shaw's famous play *Pygmalion* (Grosvenor and Wesson 1997, 14–23).

The Bell family moved from Scotland to Ontario in Canada in 1870. In 1871 A. G. Bell moved again this time to Boston where he began teaching his father's system of visible speech at the Boston School for Deaf Mutes. Over the next couple of years he helped found and edited a periodical called *Visible Speech Pioneer* and established himself at Boston University. Bell was also keen to establish his financial independence and began to mix with his interests in teaching the deaf an interest in the potentially lucrative business of building improvements to the telegraph. Between 1872 and 1874, Bell devoted energy to developing harmonic telegraph devices. In early 1874, he enquired about taking out a patent caveat on some his ideas for telegraph devices, but was informed as he was an "alien" he could only file a fully fledged patent something he was not yet ready to do. He made similar inquiries to the British Patent Office but would also encounter difficulties here, as he was not a current resident and they were unable to promise protection in his absence. Momentarily discouraged he refocused his energy on speech and sound and acoustics (Grosvenor and Wesson 1997, 45–46).

One of Bell's preoccupations was to think of ways that the deaf maybe able to "see" speech. He became especially interested in a device called the ear phonautograph. This was one of a number of devices that could produce visible images of the patterns of waves made during speech (another was the so-called manometric flame, a voice-modulated gas jet). It was hoped that the deaf could speak into such devices and gain visual feedback of the sounds they were producing—and that this would help them to develop speech. With shades of Mary Shelley's *Frankenstein*, Bell constructed a ear phonautograph in 1874. It was built using real bones from a human ear, which were mounted on a wooden frame. In response to speech the bones vibrated and a brush attached to the bones traced a pattern on a piece of smoked glass that could be rolled back and forth. Bell began to think of linking this work with his work on the harmonic telegraph: If speech could be converted into visible patterns then why not convert the vibrations of speech into an electrical current that could then be reproduced as sound (Gorman 1994; Grosvenor and Wesson 1997, 47). On November 23, Bell wrote to his family "it was a neck and neck race between Mr Gray and myself who shall complete our apparatus first [harmonic telegraph]. He has the advantage over me in being a practical electrician—but I have reason to believe that I am better acquainted with the phenomena of sound than he is—so I have an advantage there" (quoted in Grosvenor and Wesson 1997, 49).

Bell's emerging enthusiasm must have rubbed off on others as he secured funding to continue his work on the harmonic telegraph from a Boston attorney, Gardiner Greene Hubbard, and a Salem businessman, George Sanders. Both had deaf children taught by Bell. The three signed an agreement in February 1875: in return for providing financial backing they would have equal shares in any patents that Bell developed. During 1875 Bell further consolidated by securing the assistance of a skilled machinist Thomas A. Watson. To try to boost his electrical knowledge he visited Joseph Henry on March 1, 1875 at the Smithsonian. Henry told him that building a working telephone was possible and showed him a telephone-like device that had been developed inspired by the ideas of the German inventor Michael Reiss (Bruce 1973, 140). Buoyed by this visit, and in response to further experimental work carried out over the ensuing months, Bell's confidence is clear as he is said to have told Watson in a much quoted phrase: "If I can get a mechanism which will make a current of electricity vary its intensity as the air varies its density when a sound passes through it, I can telegraph any sound, even the sound of speech" (quoted in Bruce 1973, 144).

By the middle of 1875, Bell was working with the mechanism of a set of tuned reeds in a transmitter and receiver that allowed multiple messages to be sent and received at the same time. On June 2, 1875, Bell was experimenting with these reed devices with his assistant Watson. He set up three multiple telegraph stations and each had three tuned reed relays. He hoped to observe the effects on various tuned reeds along the relay as other individual reeds were plucked. At the third station, one of the reeds was stuck. Watson plucked it free so it could vibrate, as it was meant to, in response to the signal being sent down the line from its matching reed at the source. Bell had expected to hear a simple tune sound but also heard the sound of the reed being plucked, and also, although faint, a variety of complex vocal-like overtones. Rather than dismiss this as troublesome noise and a problem to be solved, Bell excitedly drew an alternative conclusion: that as little as one reed was capable, when the vibrations were being transmitted along an electrical circuit with a continuous current, to be able to receive and transmit complex voice-like tones (Bruce 1973, 146–147). Some other inventors, speculating on the possibilities of the transmission of voice electrically, had envisaged the need to produce multiple different tones to correspond to the different frequencies of the voice: a daunting project. Rather than merely transmitting an on–off pulse or single tone, by being stuck, Bell's reed, had transmitted a more complex sound.

Bell was aware that he would still need to do considerable work to develop ways to refine transmitters and receivers, and that work would be needed to increase the volume of the signal. But the moment was a defining one. While it is important to be cautious when reading recollections of the moments that may be "reshaped" with the benefit of hindsight, Watson, praising Bell, recalled this as the moment of the birth of the speaking telephone "the right man had the mechanism at his ear during that fleeting moment, and instantly recognized the transcendent importance of that faint sound thus electrically transmitted. The shout I heard and his excited rush into my room were the result of that recognition. The speaking telephone was born at that moment" (quoted in Gorman 1994, 25).

By July, Bell and Watson had begun to experiment with various membrane transmitters and receivers that further improved the volume of "voice-like" sounds. Bell and Watson's basic rationale was to use a membrane that through the effect of the air pressure produced by the voice cause a small steel armature to vibrate in front of the pole of an electromagnet. The magnet would move in response to the armature causing an unbroken undulating current to flow down a wire. This current would then activate a reed receiver that would reproduce something like the original sound (Grosvenor and Wesson 1997, 62).

Confident of making further progress, in early 1876, Bell decided to lodge a patent for "improvements to the telegraph," and the "magneto-electric telephone," including a speaking telegraph. The U.S. Patent Office had, in 1870, dropped its requirement for a working model to accompany a patent application. So Bell was able to apply for a patent even though the details for an actual working model were still extremely sketchy. Bell hoped to have his patent recognized in Britain as well as the United States but this meant it would need to be filed in Britain first. An agent was hired to do this, but after a communication breakdown the process became delayed. Hubbard became impatient and directed his lawyers to file Bell's patent in the United States anyway. They did this on February 14, 1876, just a few hours before Gray filed his caveat. Bell's patent was formally

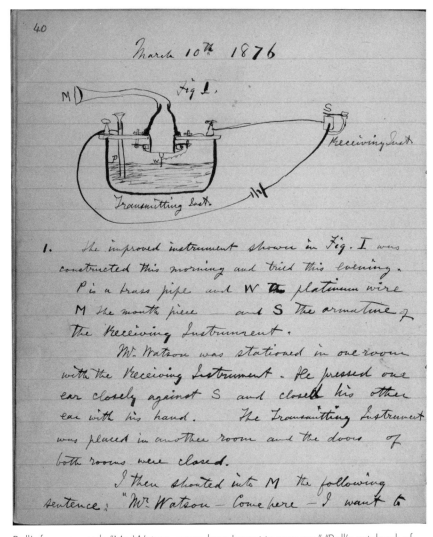

Bell's famous words "Mr. Watson, come here I want to see you." "Bell's notebook of March 10, 1876." Courtesy of the Library of Congress.

issued on March 3, 1876. This patent, U.S. Number 174,465, is possibly the most financially valuable ever issued (Lubar 1993, 122). As was noted earlier, the strange coincidence of Bell's patent and Gray's caveat arriving on the same day had not gone unnoticed even though the office favored Bell's application because it had arrived earlier and was for a full patent not just a caveat.

Like Gray's advisors, Bell's early financial backers were initially much more interested in the potential benefits of Bell's work in improving the harmonic telegraph and encouraged him to keep devote his energies to this task. But Bell's imagination had been "fired up." So in spite of this advice he would continue to devote most of his energy to his passion to develop the telephone. He quickly returned to his telephone work performing experiments similar to those suggested in Gray's caveat. A mouthpiece would transmit vibrations from the voice to a diaphragm linked to a needle dipped into a dish of acid water that changed the electrical resistance of the line. On March 10, 1876, Bell would record in his notebooks that he successfully shouted down the mouthpiece a message to his assistant (supposedly he had spilt some acid on himself) "Mr Watson, come here, I want to see you" (Lubar 1993, 122).

"ON THE MARGIN": BELL'S DEBT TO GRAY?

There has been some speculation on how much this work was influenced by Bell being able to gain access to ideas that appeared in Gray's caveat. Bell had not previously used such devices, although in the margin of his patent 174,465 he had made mention of the idea of variable liquid resistance. Bell claimed that the text that had been added in the margin had been added prior to it being submitted, but the fact that these notes appear in the margin has lead to speculation that either Bell or his lawyers had seen or been told of the contents of Gray's caveat, and that Bell had been provided with an opportunity to incorporate reference to these ideas into the margin of the text (Winston 1998, 47). These claims are not implausible, as Bell may have been provided with an opportunity to do this on a visit he made to the Patent Office on February 29, 1876. Here he met with Zenas F. Wilber, the examiner for the Patent Office, and discussed the relationship between his current application and a prior harmonic telegraph patent application he had submitted (Grosvenor and Wesson 1997, 65–66). To add to the intrigue, in later legal documents, in 1885 and 1886, Wilber confessed that he had illegally allowed Bell to actually view Gray's caveat. Bell's biographer Robert Bruce discounts Wilber's reliability as a witness suggesting he was either bribed, drunk, or both (Bruce 1973, 278).

Aside from whether or not Bell drew from Gray's ideas, using a liquid to change the resistance of the line proved unwieldy, and Bell and Watson worked hard to replace it using systems that relied on permanent magnets, electromagnetic transmitters, metal diaphragms, and on the voice producing a weak electromagnetic current.

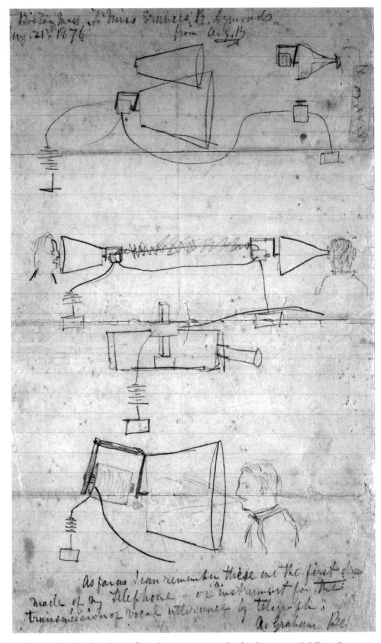

Bell's notebook sketches of early experimental telephones in 1876. Courtesy of the Library of Congress.

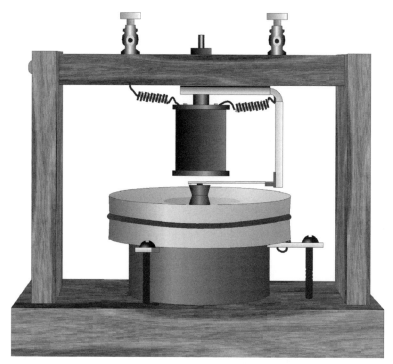

Bell's Gallow's telephone. Courtesy Robert B.K. Brown, 2006.

During 1876, a practical telephone began to take shape and would soon be ready for public display. The infant telephone still had teething problems, the signal was rather unclear, and there were uncertainties about how it would be best put to use. The turbulent infancy of the telephone, Bell's efforts to promote it, the battles over Bell's patents, the era where the telephone moved from being an "electric toy" to the device around which a whole industry would be based will be the subjects of the next chapter.

3

From Electric Toy to Business Tool, 1876–1893

GARDINER GREEN HUBBARD, PROMOTING THE TELEPHONE

Sanders and Hubbard had provided financial support to Bell in anticipation that the development of the harmonic telegraph would bring huge financial rewards to whoever could secure the first patents. Hubbard's political views provided a further incentive: he was a trenchant critic of the existing structure of the telegraph industry. One of the industries most conspicuous features was the emergence by 1866 of the powerful telegraph monopoly Western Union. Its new president from 1867, William Orton, would further consolidate this monopoly and would shape its market strategy to be geared to sending short business messages between major cities.

Hubbard launched numerous public political attacks on the Western Union monopoly. He cautioned that the monopoly would be able to raise prices without restraint, would discourage the development of technology to serve wider users, and would have too much access to market information, private business messages, and news. This access allowed them to potentially influence markets and interfere with freedom of the press. Hubbard argued that an unchecked Western Union was a threat to American democracy and that to answer this threat the telegraph system needed to be modified to accommodate a wider market than business communication, and be more open to the "day-to-day" communication needs of the average citizen. Various "practical" suggestions were made to the Congress, including letting the Post Office construct its own telegraph lines and the government underwrite a second telegraph network to provide competition. Hubbard's postal telegraph system was never supported by the Congress but his ideology contained some of the seeds for the later idea which would run through the history of the telephone, of universal service: that telephones should be as widely available as possible so as to encourage, through better communication, national unity and democracy (Carlson 2001, 25–56).

While Hubbard was initially more enthusiastic for Bell to stick with his work on the harmonic telegraph, he helped Bell to arrange a number of lectures and demonstrations of the "telephone"—still really in something of an experimental form during 1876. Demonstrations included one on May 10, 1876, to a meeting of the American Academy of Arts, where he cited some 30 articles from American, English, and French scholarly journals concerning "the acoustic effects of magnetic processes" (Flichy 1995, 83); another on May 25 to the Massachusetts Institute of Technology; and, most importantly, on June 25 to the Centennial Exhibition. Bell's demonstration became a significant event at the exhibition. Bell's first public words down the telephone were from the soliloquy of Shakespeare's *Hamlet*, "'to be or not to be.'" In response Dom Pedro, the Emperor of Brazil, is widely reported to have excitedly shouted "My god it talks" (Grosvenor and Wesson 1997, 72–73; Lubar 1993, 122).

Bell's device quickly became well known, but it would still take some time and effort for it to move beyond being seen by most serious "telegraph men" as more than just an interesting electric toy. In the wake of the Centenary Exhibition, inspired by Bell's demonstration, Gray would out of curiosity build some telephone devices, and in March 1877 an article appeared in the *Chicago Tribune* reporting that Gray had invented the telephone. Bell hearing of Gray's tinkering and the *Tribune* article wrote an objection to Gray, who replied in part: "I do not however claim even the credit of inventing it [the telephone], as I do not believe a mere description of an idea that has never been reduced to practice—in the strict sense of that phrase—should be dignified with the name invention" (quoted in Grosvenor and Wesson 1997, 79). Gray's diplomatic and rather reticent response is consistent with the general belief that the telephone was still a scientific curiosity. Gray's views would change as the significance of the telephone quickly became apparent. This correspondence with Bell came back to haunt Gray in later legal proceedings.

WESTERN UNION TURN DOWN THE RIGHTS TO BELL'S PATENTS

While these demonstrations received considerable interest, Bell and Watson encountered numerous practical challenges. Their phones of early 1877 relied on the same hardware to provide the transmitter and receiver. Callers spoke into a box that looked a little like an old-fashioned camera. Callers spoke loudly into a "tube-like" opening and then turned their heads to listen for the reply. This was obviously an uncomfortable arrangement, the quality of transmission was poor, forms of switching for multiple calls were still to be developed, and future users had to be convinced the device would be worthwhile. It is possible to speculate that awareness of these technical difficulties may have been one of the factors (among others) which encouraged Hubbard and Sanders in late 1876 to attempt sell their telephone patent rights to Western Union for $100,000. Hubbard may have hoped that by embracing such new patents and new technology Western Union may have been encouraged to spread its services to wider users and not just focus on its business clients. He may have also simply thought the money could provide some financial security for Bell who would soon be his "son-in-law": Bell had recently become engaged to Hubbard's deaf daughter Mabel (Carlson 2001, 40).

In what, with the benefit of hindsight, appears a strange decision Western Union turned down the offer. Orton reportedly quipped, "what use could this company make of an electrical toy" (Aronsen 1977, 16). He did not even see fit to buy the patent to put a hold on the development of Bell's telephone. Some have explained Orton's decision as an example of the reasoning of a spoilt monopolist: What need would he have for the unproved telephone when he ran a lucrative telegraph business? Alternatively: Was it perhaps a display of bloody-mindedness given that Hubbard had been a long-standing source of irritation to him? Orton's decision not to buy Bell's patents also has a set of more tangible explanations. Bell's telephone, as of 1876, had a significant number of technical limitations: it was still difficult to have a clear two-way conversation and, in some early demonstrations, confirmation that a voice message had been successfully transmitted required the return of a separate telegraph message. Orton was also aware of the skills of Gray and Edison. It was quite likely that they would be capable of creating better similar devices in the near future that would be able to beat Bell in any future patent disputes. Orton would in fact go on to commission Edison to start working on improvements to the telephone transmitter from the early part of 1877.

Orton's reluctance also reflected the fact that in 1876 the telegraph was a thoroughly socially entrenched successful technology. It had enjoyed considerable incremental improvement and was increasingly being applied in a number of broader contexts beyond conveying simple Morse code dispatches. Developments such as "autographic telegraphy" that sent handwritten messages, and multiplex systems that allowed an increasing number of signals to pass down lines simultaneously, were coming into use. From the second half of the 1860s telegraph exchanges, mainly servicing large banks, had also begun to appear in cities such as Philadelphia (1867) and in New York (1869). These exchanges provided the opportunity not only for telegraphic messages but also for telegraphic conversations, which hinted at the future business possibilities of the telephone (Flichy 1995, 86). The first major telephone networks would serve similar business functions and these "telegraph networks" would be later targeted by Bell as sites where the telephone system could be introduced. It was not always immediately clear, aside from its novelty appeal and scientific ingenuity, how Bell's invention would be used and how it would offer improvements over the telegraph.

It is also worth noting that Orton was also not alone in refusing to buy Bell's early patents. The British Post Office Department, who controlled the telegraph in England, also turned down options to buy them (Aronsen 1977, 19). While Bell himself appeared to have a vision of the telephone as a vehicle for "one-to-one" conversations, and a significant part of his lectures promoting the telephone alluded to these possibilities, his visions were not as immediately clear to others.

INVENTING USES FOR THE TELEPHONE

Bell's telephone promotional "road shows" now began in earnest. They offered a source of finance and publicity and offered a means to maintain the confidence of Sanders and Hubbard. One of the clever techniques that Bell and Watson often used in their demonstrations of the early telephone was to recite well-known phrases or songs. Bell was aware that in using these familiar phrases and songs the listeners would by their powers of anticipation of the familiar compensate for the poor quality of the signal. Bell would set up special telephone events: Watson would sing down a telephone from one end of town to an audience at another location such as a church hall. A New York 1877 poster announced that for a 25-cents admission they could attend: "An Entertainment of the Sunday School of Old John St. M.E. Church," including recitations, singing, and an exhibition of

Prof Bell's Speaking and Singing Telephone" (from a poster reproduced in Stern and Gwathmey 1994, 14).

Familiarizing possible future users with the telephone was also important. In a pre-telephone world it wasn't always immediately clear to people how Bell's telephone should be used. People were curious about whether the telephone spoke only in English and what form of speech would be best to use. One of Bell's first advertisements reminded the public "conversation can easily be carried on after slight practice and with occasional repetition of a word or sentence. On first listening to the Telephone, though the sound is perfectly audible, the articulation seems to be indistinct: but after a few trials the ear becomes accustomed to the peculiar sound" (cited in Lubar 1993, 125).

On July 9, 1877, Hubbard, Bell, and Sanders formed the Bell Telephone Company. Hubbard was the trustee (from this point on in the text *Bell* will be italicized when it refers to the various corporate arrangements often known as the Bell system). They initially sold licenses for the establishment of lines between businesses and between homes and offices. Individual licensees would be responsible for setting up the line themselves. Hubbard went on to pursue a variety of strategies for promoting the telephone. He suggested that the telephone could greatly increase the efficiency of the telegraph operator. Operators using Morse code could normally transmit about 15 messages a minute; with a telephone they could now transmit between 150 and 200. More novel were Hubbard's suggestions that private telephone lines linking different offices, and home and office, could be developed as a cheaper alternative to the telegraph. A small number of Bostonian entrepreneurs were sufficiently interested to purchase licenses to build private lines. Some of these efforts were inspired by burglar alarm and fire alarm networks that operated via telegraph lines.

Hubbard also promoted the telephone with the rhetoric reflecting his political vision that its adoption would challenge Western Union's monopoly, promote the emerging middle class, and nurture the American democracy. He emphasized the value of the telephone to the upper middle classes to coordinate servants, order groceries, and communicate socially (Carlson 2001, 41–43). The broader spread of the telephone beyond businesses and upper middle classes and professionals would in fact take some time. Despite its initially narrow adoption the telephone did take its place in the public imagination (amid other inventions of the new age of electricity) as a reassuring sign of optimism in American inventiveness and progress that could act as a counterweight to the economic and social uncertainties following the U.S. civil war. It is easy to forget that at the same time the telephone was being developed there was a growing pool of poor immigrants in the cities of the Northeast, the populations spreading along the western frontier faced numerous hardships and challenges, and there was major labor unrest in major industries such as the railways (Carlson 2001, 44–45; Smith 1996, 1–35).

Hubbard made a number of key business decisions that would influence the future development of the telephone. Possibly the most significant was the decision to maintain *Bell* as the exclusive builder of the telephone and that local providers of the telephone service would lease the instruments and provide the actual service under license. The system developed so that franchisees would use their own capital to rent telephones, construct the necessary switchboards and wiring, and organize subscribers. As the system grew *Bell* was able to, via licensing renewals and by setting rates and standards of service, exert an influence on the way services were delivered by franchisees. From this environment a set of so-called local but linked "*Bell* operating companies" would emerge. In a pattern that would persist well into the next century, *Bell* would control the development of the telephone by providing technical know-how, equipment, and the collection of rental fees (Fischer 1992, 36).

WESTERN UNION ENTERS THE TELEPHONE BUSINESS: PATENT WARS, 1877–1879

It did not take long, by September 1877, for Western Union to take on a short-lived but intense interest in competing with the emerging *Bell* system. Rather than buy licenses from *Bell*, they would buy patents from Edison, Gray, and other telephone inventors. In December 1877 Western Union created the American Speaking Telephone Company. Western Union began to adopt telephones designed by Edison and Gray and others in an attempt to offer competition. One important technical development would stand out as being particularly relevant to these battles: the need for improvements to the telephone transmitter. Emile Berliner (who would later work for *Bell* and contribute to the development of the phonograph) and Thomas Edison both came up with similar ideas in 1877 for improvements in telephone transmitters (Meyer 1995, 14). *Bell's* early commercial transmitters used energy in sound waves from the voice to induce an electric current in the line. This system had limitations and couldn't produce a clear loud signal that could be carried over long distances. Berliner and Edison worked on models for "contact pressure" transmitters that worked much more effectively than Bell's original designs and were the forerunner to the idea of the microphone. Berliners patent was based on "[the] method of producing in a circuit electrical undulations similar in form to sound waves by causing sound waves to vary the pressure between electrodes in constant contact, so as to strengthen and weaken the contact and thereby increase and diminish the resistance of the circuit" (quoted in Meyer 1995, 15).

Edison and Berliner had observed the way telegraph operators when sending a message over a longer distance could do it more effectively by placing more physical pressure on their keys. Edison and Berliner would engage in a patent dispute over who should get the credit for the improved transmitter. Aware of the work of Edison and others, *Bell* quickly put energy into trying to introduce their own improvements and buy the patents of other inventors (Winston 1998, 56–57). While *Bell* had already installed a large number of telephones, by mid-1878 an estimated 10,000 *Bell* instruments were in operation in the United States (Fischer 1992, 36–37) the "American Speaking Telephone Company" had the advantage of access to the hundreds of thousands of miles of telegraph lines controlled by Western Union and the steadily improving transmitters being developed by Edison.

The race to produce better transmitters was intense and in 1878 Francis Blake of Massachusetts produced yet another improved variation of the contact pressure transmitter. *Bell* bought Blake's patent and employed him. His transmitters produced sufficient quality of voice transmission to allow *Bell* to compete with Western Union. These transmitters still nevertheless, relied on making a single contact between electrodes that limited the strength of the signal. About the same time, Henry Hunnings in England, and yet again the irrepressible Edison, would address the problem of making and controlling multiple contacts by using carbon granules to fill the spaces between the electrodes in the transmitter. These granules responded to the variable pressure produced by sound waves from the voice particularly effectively. Edison's insights would influence the design of telephone transmitters for many years to follow (Meyer 1995, 14–16).

Further telephone innovations of 1878 included Thomas Watson's telephone ringer and the establishment of the first telephone exchanges: by *Bell* on January 28, 1878, in New Haven Connecticut and less than a month later, by Western Union on February 17 in San Francisco. Prior to these exchanges, lines were simply between specific individuals or organizations. February 28, 1878, also saw the birth of the first telephone directory by the New Haven District Telephone Company and the first telephone being installed in the White House for President Rutherford B. Hayes (Farley 2006). During this time a number of basic telephone designs also began to become popular. One of the first commercial telephone designs was the so-called 1878 "Butterstamp" telephone (the receiver looked like a then-popular device, the Butterstamp). This phone relied on combining the receiver and the transmitter into one unit. While still ungainly it offered

Thomas Edison would provide significant
improvements to the design of the telephone.
Portrait of Thomas Edison. Courtesy of the
Library of Congress.

an improvement on Bell's first "camera box" designs, in that at least the
combined mouthpiece and receiver could be lifted off the bulky transmitter
unit, which was attached to the wall. The Butterstamp was replaced by a
variation in design which had a second transmitter and receiver (Stern and
Gwathmey 1994, 33–34).

With these emerging technical improvements the stakes involved in
controlling the emerging telephone business would become more intense.
In 1878, *Bell* initiated legal action against Western Union claiming that
they were infringing on *Bells* patents. In all, *Bell* would lodge some 600
notices of interference with its patents. A. G. Bell would find himself in
court, pitted against rival inventors such as Gray, who by now was becoming
increasingly embittered. It is interesting to compare the tenor of a report on
a reception given to Gray in November 1878 by the residents of Highland
Park with Gray's earlier reticence of just over a year before: "It should be
known that Elisha Gray a resident of Highland Park, and a gentleman of

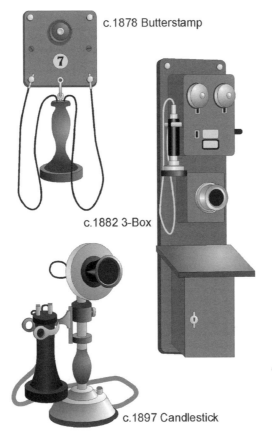

Some common early telephone designs. Courtesy
Robert B.K. Brown, 2006.

superior scientific attainments ... is the individual to whom beyond all
doubt, the world is indebted for the original invention of the speaking and
musical telephone" (quoted in Stern and Gwathmey 1994, 10).

The telephone's infancy was proving to be an unsettled one, shaped by a
pattern of ongoing micro-inventions, initiated against a backdrop of desires
not only to improve the way the telephone functioned but also how it
might contribute to the ongoing strategies involved in legal wrangling over
patent rights. While *Bell's* claims had proved to be quite resilient when faced
with legal challenges, both sides saw the financial merits of a compromise
and by the end of 1879 *Bell* and Western Union would settle their dispute
(although various claims by rival inventors would still linger on). *Bell* would
have all patent rights on telephone instruments and in return would not
enter into telegraph services. It would also pay Western Union 20 percent
of *Bell* rentals for the life of *Bell's* patents (17 years) and grant some Western

Union interests in some local *Bell* companies. *Bell's* key patents would cease to be effective in 1893–1894 (Grosvenor and Wesson 1997, 96–98). While the original agreement contained clauses which suggested that *Bell* focus on developing a market for personal conversations, and limit its competition with Western Union's interest in general business messages, in reality it would be in these commercial areas, duplicating and modifying commercial communication demands previously satisfied by the Telegraph that *Bell* would first expand its business (Flichy 1995, 86).

Bell was left after this settlement, in early 1880, with approximately 60,000 subscribers. By the beginning of 1881, *Bell* controlled 132,692 telephones and held a virtual monopoly (which persisted till 1893–1894) over the U.S. telephone business (Grosvenor and Wesson 1997, 122). The year 1880 also saw *National Bell* become *American Bell* and the formal withdraw of Hubbard and A. G. Bell from the company with significant fortunes.

A. G. Bell went on to lead a fascinating life never tiring of inventing, experimenting with ideas such as the phonograph, a "photophone" where light was used to convey sound, designing "flying machines," or "aerodromes" as he called them, becoming President of the National Geographic Association, taking over the management of the journal *Science*, encouraging the adoption of Montessori educational models, and supporting women's rights to the vote. His biographers paint the picture of a man who led an extremely rich and virtuous life. This would appear to be largely true although his main biographers fail to mention that as a man of his times he also promoted some ideas that would be less palatable to contemporary sensibilities such as eugenics (Sterling 1995, 36). Eugenics was the "science" of selective human breeding. Through his support of eugenics—despite devoting large parts of his life to helping the deaf—he believed that those who were congenitally deaf should be discouraged from having children: his own wife, and mother of his children, Mabel, acquired her deafness as the result of illness as a child, so it didn't carry to her offspring.

FURTHER CHALLENGES AND INNOVATIONS

The spread of telephones continued to create technical challenges to *Bell*. Stray electrical currents spilling from streetcar power lines and other telephone lines meant that poor reception and tangled masses of wires were both common, and popular models of telephones relied on batteries and other devices located in the home that were frequently leaky and unreliable. *Bell* responded to these problems by moving toward providing power from switch boards, rather than from each phone, and developing new forms of wiring most significantly taking out a patent in 1881 for a two-wire metallic circuit which would gradually replace the original poorly grounded, noisy single-line circuits. *Bell* businesses fostered research and acquired as many relevant patents to support incremental improvements in telephone design as possible (Grosvenor and Wesson 1997, 121–124). The role of *Bell* in encouraging innovations to telephone technology is something that was to become increasingly important in the future.

THE TELEPHONE AS A NETWORK

As noted in earlier discussion many perceived that the telephone would primarily offer a system of paired stations. The development of the telephone primarily for "point-to-point" communication was not always the only obvious possibility. The telephone also offered the possibility of broadcasts and "party-lines" where a number of callers could converse simultaneously. Foreshadowing the later emergence of radio there were some suggestions that telephones might be used to broadcast music and news. Vice President E. J. Hall of the American Telephone and Telegraph Company commented at an 1890 Detroit convention of the telephone industry: "More wonderful still is a scheme which we now have on foot, which looks to providing music on tap at certain times every day, especially at meal times. The scheme is to have a fine band perform the choicest music, gather the sound waves, and distribute them to any number of subscribers. Thus a family, club or hotel may be regaled with the choicest airs from their favorite operas while enjoying the evening meal, and the effect will be real and enjoyable as though the performers were actually present in the apartment" (quoted in Briggs 1977, 43–44). Despite the telephone's potential to be a public or broadcast technology the commercial potentials of networks of paired telephones had already been foreshadowed by the successful uses of telegraph networks for businesses such as banks, and for subscribers to things like fire and security alarm systems. These uses of the telephone required the development of telephone exchanges to help coordinate the huge number of lines that were growing between subscribers. The first telephone exchanges relied on manual switching. An operator would receive a verbal request for a connection and then physically plug one line into a switchboard so it would connect with another. As exchanges grew a number of operators would work in close proximity, any individual operator would only be responsible for a designated number of incoming calls but would be able to manually reach the connections of all subscribers to the exchange so as to complete connections.

WOMEN AND THE MANUAL SWITCH

These first manual switchboards, following the example of the telegraph industry, used the labor of young men. A visitor to an early telephone exchange described the scene "The racket is almost deafening. Boys are rushing madly hither and thither, while others are putting in or taking out pegs from a central framework as if they were lunatics engaged in a game of fox and geese" (quoted in Lubar 1993, 126). Telephone exchange boys were quickly replaced by young women. Why young women came to dominate the task of the telephone operator and the possible influence this may have had on the development of automatic telephone switching technology are questions of some historical interest, and it is worth taking a little time to explore them in more depth. At a superficial level the story of the development of telephone switching is the story of inventors endeavoring to solve the problem of more effectively switching telephone calls from one caller to the next. Human operators manually doing this constituted an added cost to the telephone system, a possible privacy problem, a source of possible human error, and presented limits to the number of calls that an exchange could handle. In this abridged version of history, inventors gradually solved these technical problems and automatic switching would become an inevitable part of the destiny of the telephone. The more detailed story of automatic switching is much more interesting and revealing of the time and culture in which the telephone developed. One of the key factors that shaped this phase of the development of the telephone was in fact gender: the role fulfilled by female telephone switch operators.

The new manual telephone exchanges required fast accurate switching, considerable discipline, but most importantly the operator also needed to briefly speak and interact with customers. Managers quickly came to view young women as more articulate, polite, and more likely to follow instructions, than male operators. Male operators were also more likely to be abused by disgruntled customers when technology malfunctioned (as it often did) and to become industrially organized (female telephone operators would not organize industrially till 1920). A "telephone man" of 1881 extolled the virtues of employing young women to work the new telephone exchanges "I would like to say right here, I've been asked by Mr. Sabin what our experience has been with the young ladies' help; the service is very much superior to that of boys and men. They are steadier, do not drink beer and are always on hand" (quoted in Winston 1998, 248).

Some historians have suggested that one of the integral roles played by women manual exchange operators was their provision of "personal service." Personal service effectively involved young women playing the

Women working at manual switches would dress neatly as was in keeping with their task of not only efficiently transferring calls but also providing polite personal service to callers. "Women switchboard operators." Courtesy of the Library of Congress.

role in the telephone system parallel to the female maid, something which fitted in with *Bell's* belief in its early years of operation, that its main market of users would be businesses and the upper middle classes. In the early years of operation, the unreliability of the technology meant that *Bell* also looked to "auxiliary services" that could be supplied by the polite female operator as a way of keeping customers satisfied. Early services included messenger services and weather and sports results offered at minimum or no charge. For the *Bell* manager, "the personalization of service [meant] . . . a service that is not only as nearly perfect technically as possible, but as pleasing as possible to the telephone user" (quoted in Green 1995, 918).

Exchange operators were meant to be efficient, confidential and courteous, educated, native born, without strong accent, white, and in most respects resemble the ideal domestic servant. *Bell* was able to draw upon a relatively large pool of "appropriate" female labor, as there was an abundance of educated young women from middle and aspiring working class families who upon the completion of high school found it difficult to get work outside service and clerical occupations. The emerging telephone industry also constituted a socially respectable niche for female labor. Operators were expected to be young, leave upon marriage, and their employment track was fairly limited, although some moved to the rank of supervisor. As exchanges grew, training became important as operators had to learn to smoothly coordinate their own movements and reactions with increasingly complex, often unwieldy, switchboard technology. One of *Bell's* first female switchboard operators, Katherine Schmitt, suggested that the operators "must be a paragon of perfection, a kind of human machine" (quoted in Lipartito 1994, 1088).

While the customer never saw the operator, the idea of telephone operators being part of polite respectable culture was strongly cultivated. Across most of the Bell system, neat dress codes applied and managers took an interest in their operators personal and family life, sending on occasions "medical matrons" to report on their family arrangements. One operator recalled how she wore for her first day on the job in New York in 1881 "a garnet dress of cashmere, the waist was very well boned . . . and a turnover white linen collar fastened with a bow" (quoted in Maddox 1977, 268).

THE STROWGER SWITCH

During the time that female telephone operators became standard for *Bell*, an alternative model for operating exchanges, the automatic switch, had been patented and many inventors had put their minds to the task of removing the need for telephone operators. Between 1879 and 1898 more than 86 new automatic switching systems were patented and offered for sale to *Bell*. Some limited automatic systems had been developed to satisfy contexts such as small towns where not enough calls could be made to justify the salary of an operator. In 1885 engineers installed E. T. Gilland's so-called "village system" which had a capacity of 40 customers who had no need to go through a central human operator. This was installed in Leicester in Massachusetts (Green 1995, 926). Gilland's switch would serve as one of the inspirations for the development of what would later become the most important automatic system, the Strowger switch. This system was based on ideas patented by Almon Strowger in 1891. Strowger was a rather enigmatic figure, born in 1839, near Rochester, New York. He had served in the Civil War, been a schoolteacher, and then an undertaker in Kansas where he developed his ideas for an automatic telephone switching system. The oft repeated story, that forms part of the mythology of the history of the telephone, notes that Strowger apparently made regular complaints to *Bell* about his poor telephone service, and was less than impressed with the "personal service" of local telephone switch operators who he believed were deliberately directing business away from him to a rival undertaker. This provided him with the motivation to think of ways of improving privacy and replacing untrustworthy manual operators. Strowger's nephew later claimed that Almon had stolen the design from his own brother Walter, who actually carried out most of the development of the system with the assistance of an associate Joseph Harris, and would go on to help develop one of the first companies that marketed automatic switching, Automatic Electric (Lipartito 1994, 1095).

Bell was aware and interested in the idea of automatic switching, but seeing it as a solution to switching calls in small towns was not the same as seeing it as solution to servicing larger numbers of customers in larger urban areas. *Bell's* patent attorney, Thomas D. Lockwood, stated, "No intelligent person who had experience in telephone exchange work ... would seriously entertain such a proposition as using automatics for large exchanges" (Green 1994, 927). Through the lens of *Bell* advisors, who were accustomed to manual exchanges, automatic switching was complex, required expensive wiring that needed to be maintained by skilled labor and was vulnerable to breakdown. The use of automatic exchanges might also lose customers if users needed to perform a number of operations themselves. It is easy to forget that in the early years of the telephone, instructions were issued about just how to speak on the telephone, let alone perform technical tasks such as dialing. The idea that the user should need to know numbers and perform multiple operations on their telephone would take time to be viewed as fully legitimate. In all, female-operated manual switchboards fitted in better with many managers' perceptions of technical efficacy, the role of the telephone user, and that part of the business of running and developing telephone systems was to provide a particular standard of "personal service." Even as automatic switching became increasingly cheaper and addressed growing concerns about privacy, introducing it may have inspired other changes to the telephone system managers were uncomfortable with.

THEODORE N. VAILS'S FIRST TERM, 1878–1887

Possibly the most important figure in the promotion of the telephone after A. G. Bell and Hubbard was Theodore N. Vail (interestingly a nephew of Alfred Vail who had assisted Morse). Theodore Vail had been the manager of the American Railway Postal Network before being offered the role of general manager and president (a position he filled from 1878 to 1887) by *Bell's* financiers. He would return in 1907. Vail would be a key figure in the early years of the telephone, in some ways putting into practice the ideas foreshadowed by Hubbard. He would become an even more central figure in later eras of the development of the telephone as a mature technological system. Vail began to create a business model for *Bell* that would be composed of regional companies providing local services, a manufacturing wing, and a long-distance service. Western Electric (founded by Elisha Gray and Enos Barton [*Bell* would take them over in 1881]) would form the manufacturing wing and long-distance service would be performed by AT&T, created on February 28, 1885. Vail believed that AT&T would help maintain *Bell's* virtual monopoly when its key early patents expired in 1893 and 1894. This would occur as AT&T would only connect their long-distance service with local companies working under *Bell's* licenses (Farley 2006).

These strategies of continuous investment in incremental improvement to the quality of telephone technology and improvement in service were pursued ahead of working on cutting costs for users. So while the early era of the telephone saw steady technological improvements, there were regular complaints by early telephone users that costs were too high and there is evidence that there were various disagreements and discussions about what forms of pricing and services the emerging industry should promote and adopt. In one case customers became so irate at *Bell's* pricing policies that there was a boycott of *Bell* in Rochester from November 1886 to May 12, 1888, with the Rochester Common Council threatening to cut down the telephone poles on Main Street (Grosvenor and Wesson 1997, 130–132).

Because many of those involved in the early telephone industry had previously been involved in the telegraph industry they assumed that the telephone would primarily function as a business tool. Their belief that the telephone would offer a replacement for the telegraph in business contexts was reinforced by the fact that businesses were the dominant users in the telephone's early years. Some professional groups, such as physicians, were also well represented. Residences more generally were less well represented. In an analysis of telephone usage in 1879 in Pittsburg, of 300 telephone lines 294 were owned by professionals with 6 being used by entrepreneurs who used them to maintain links between home and factories (Aaronsen 1977, 27–28; Flichy 1995, 86).

Vail took an interest in trying to expand *Bell* into broader residential markets, but one of the main constraints was price: residential rates would have to be reduced. *Bell's* financial backers had little interest in this, and rhetoric aside, little headway was made. Even in the early years after competition was introduced in 1896 a New York phone service was $20 a month relative to an average worker's wage of $38.50 a month (Fischer 1992, 48–49). Vail resigned from *Bell* in September 1887. He was unhappy with *Bell's* neglect in providing services to various rural areas and their inflexibility in applying continuously high charges: "We have a duty to the public at large to make our service as good as possible, and as universal as possible, and that earnings should be used not only to reward investors for their investment but also to accomplish these objectives" (quoted in Farley 2006). This notion of "universal service" was to become a central theme that has come to be repeated throughout the life of the telephone. Vail would return to shape the telephone system in 1907. Between 1880 and 1893 telephone use grew steadily in the United States to about 260,000 from 60,000 (Fischer 1992, 46). But most of this was steady growth in business use. Hubbard's vision of a broader set of telephone users and reality behind Vail's rhetoric of "universal service" would still be some time in coming (Mueller 1997).

4

Expansion, Competition, and the Remaking of the Bell Monopoly, 1893–1918

◆

In February 1893 the *Chicago Evening Journal* editorialized: "The American Bell Company has for years been forearming itself against the ides of March, 1893. By purchase and otherwise it has acquired the patent right of almost every practicable telephone transmitter and receiver. Hundreds of such patent rights, through which alone successful competition might come, lie securely locked in the safes of the big parent Bell Company, never to see the light of day, it may be unless the company adopts them for its own apparatus" (quoted in Grosvenor and Wesson 1997, 163).

When *Bell's* key telephone patents expired in 1893–1894, there was a burst of activity as new companies entered into the telephone business. *Bell* controlled the technology and companies starting from scratch had to not just offer telephones but a system of wires, trained technicians, exchanges, and operators. Despite *Bell's* dominance it had placed little energy into developing its activities outside the key business and population centres. So there were significant opportunities for independents to grow in rural areas. By 1894 there were 87 independent companies. This number grew by 1902 to more than 6,000 (Fischer 1992, 43). Many of these were tiny operations, so-called "mutual" companies operated by farmers. In keeping with the rural embrace and adaptation of other forms of mechanization farmers would often purchase their basic telephone equipment via mail order. Because they had fewer problems with other forms of electrical interference such as electric wires and streetcars than those living in cities, farmers often strung their own cheap single wire lines using their fences. In cities a smaller but still significant number of independents began to grow. On the whole, independent services tended to offer inferior equipment and limited area of service but tapped into broader markets by being cheaper.

Independents did occasionally offer new better technology and service. The "Automatic Electric Company" (this was the company that had emerged from Almon Strowger's patent), for instance, expanded by offering alternatives to *Bell's* manual switchboards and constructed dial telephones. In 1905 The Automatic Electric Company could boast sales of 8,000 dial phones in Chicago and 19,300 in Los Angeles. Other firms which challenged *Bell* with claims of better technology included Stromberg and Carlson, originally based in Chicago in 1894, but moving to Rochester in 1899 and becoming the Home Telephone Company. In less than a decade from the expiry of *Bell's* patents, *Bell* temporarily lost its grip as the telephone monopoly (Farley 2006).

By 1902, 45 percent of communities with more than 4,000 people had two telephone services. By 1903 independents managed more phones than *Bell*: 2,000,000 to Bell's 1,278,000 (but with *Bell* still controlling more than two-thirds of the wires). In these settings, *Bell* had continued to be the predominant service to business and the upper-middle classes, to a significant extent because it offered superior access and service involving long lines. The failure of independent lines to be able to link to the larger preexisting *Bell* system nevertheless constituted an ongoing problem. Some combined in 1897 to form the National Association of Independent Telephone Exchanges with the goal of facilitating the development of long lines. By 1905 they had developed some, wider better integrated services, mainly in the east, but they lacked *Bell's* capacity to link together major business centres which ultimately limited their profits (Winston 1998, 250–252).

This era of competition was marked by a significant cut in the cost of telephones, and also revenues per phone for *Bell*. In 1895 *Bell* produced operating revenue of $88 per telephone; by 1907, this was $43. In 1882, city telephone rates were around $100 a year in Chicago, Philadelphia, and Boston, and $150 in New York; by 1907, some rural areas were paying as little as $12 a year. Wider access was also facilitated by the growth of the pay phone, the first pay phone had been introduced in Springfield, Massachusetts, in 1883, after initial slow growth; by 1902, there were 81,000 in the United States. The desire to spread markets also encouraged some firms such as the New York Phone Company in 1896 to offer metered service encouraging lower rates for residences relative to businesses (de Sola Pool 1983, 22).

Like the earlier *Bell* monopoly era, frequent complaints about the quality of service persisted. Independents often lacked the capital to maintain equipment and service and many made short-term profits through corporate manipulations before going in to the hands of receivers. Further, initial promises of the offer of cheaper rates to compete with *Bell* were often unsustainable. Independents also were faced on occasion with *Bell* engaging in ruthless business practices. A good example was *Bell's* attempt to thwart independents attempting to capitalize on using new improved switching technology. In 1897, Milo Gifford Kellogg (he had previously helped develop switchboards for Western Electric) established the Kellogg Switchboard and Supply Company. His switchboards offered a much greater capacity than the boards used by *Bell*. These boards were becoming popular among some of the larger independents. The threat of success of Kellogg made *Bell* anxious. The then president of AT&T, Fredrick Fish approved of a plan to secretly buy out the Kellog Switchboard and Supply Company. Kellogg would keep on selling their switchboards to independent companies. *Bell* would then, after the independents had installed the boards, file a patent suit against Kellogg, which being under *Bell* control they would deliberately lose. The independents would then be forced to pull out their new Kellogg switchboards and be financially ruined. The scam was exposed and the *Bell* system's already shaky reputation was tarnished further (Grosvenor and Wesson 1997, 167). This would not be the only time *Bell* would be accused of engaging in unethical strategies to limit the expansion of independent companies into its main markets. The Peoples Telephone Company in New York, for instance, had complained that *Bell* had impeded them by preventing their access to subways which were needed to duct wires: *Bell* who held stock in the subway operations, claimed that the subways couldn't accommodate further wires (Winston 1998, 250). Despite its turbulence and uncertainties the era of telephone competition between 1893 and 1907 saw a massive growth in the number of telephones at a compound rate of 23 percent per capita (Fischer 1988, 36).

VAIL'S SECOND TERM: "ONE SYSTEM, ONE POLICY, UNIVERSAL SERVICE," 1907–1919

Bell's financial backers had become increasingly anxious at the challenges that had been thrown up by the growth in independent phone companies.

Bell had been overextending itself financially to attempt to maintain its powerful position, had developed a bad public reputation in relation to its business strategies, and its political position was vulnerable to threats of antitrust regulation. The factors made the *Bell* appear economically tenuous and in 1907 it had been unable to sell its bonds. The financial group run by the powerful, financially and politically aggressive businessman J. P. Morgan, owned $90 million of unsold AT&T bonds. This lead to the J. P. Morgan-led banking group taking control of the firm and reemploy Vail with the task of reaffirming the dominance of the *Bell* system. With Morgan's financial support numerous independents were bought up and integrated into the *Bell* system and 30 percent of the stock in Western Union was acquired in 1909. Vail revisited his earlier strategies of ensuring that its subsidiary Western Electric would become the main producer of telephone equipment in the United States and for AT&T to dominate long line service (Farley 2006).

From 1908, Vail also began advertising nationally the slogan that would later become famous: "One System, One Policy, Universal Service." The ideas of "universal" carried multiple meanings; universal in a sociological sense: the extension of telephone access to all households; universal in a technological sense: that the telephone system should be standardized; and universal in the spatial sense: that it wasn't geographically limited. The later two senses of universal were possibly the most significant in practice to *Bell* in this era. In a 1911 *Bell* system advertisement, AT&T's superior capacity to offer long-distance calls forms the central argument for the necessity of universal service. "Telephone users make more local than long distance calls yet to each user comes the vital demand for distant communication. No individual can escape this necessity. It comes to all and cannot be foreseen. No community can afford to surround itself with a sound-proof Chinese Wall and risk telephone isolation. . . . Each telephone subscriber, each community, each State demands to be the center of a talking circle which shall be large enough to include all possible needs of intercommunication. In response to this universal demand the Bell system is clearing the way for universal service"(advertisement reproduced in Mueller 1997, 102).

Vail's advertising campaign and promotion of universal service became an important way of trying to rebuild *Bell's* public image and counter ideas of dual service. Vail's promised that profit and public service could coexist: "With a large population with large potentialities, the experience of all industrial and utility enterprises has been that it adds to the permanency and undisturbed enjoyment of business, as well as to the profits, if the prices are put at such a point as will create a maximum consumption at a small percentage of profits" (Winston 1998, 256). Vail also declared that he was not an opponent of government regulation as long as it was "independent, intelligent, considerate, thorough, and just" (Winston 1998, 257).

Vail's strategies proved successful and by 1912, 83 percent of independents were connected to Bells wires. Bell's reassertion of its monopoly continued to raise the ire of independents who persistently appealed to antitrust laws, greater government regulation, and "for protection against outrageous methods of warfare which are detrimental to the public welfare"(Winston 1998, 256). Other lobbies didn't believe that competition was the answer, instead suggesting that, as was the case in most other nations, telephones would be run best by the government via the Post Office.

There had been a growth in public opposition to the power of monopolistic corporation from the turn of the century. In 1911, as an outcome of an important antitrust case, *United States v. Standard Oil*, John D. Rockefeller had been compelled to break up his business interests (Faulhaber 1987, 5). In January 1913 the Justice Department informed Vail that the Bell system was bordering on breaching the Sherman Antitrust Act. Rather than risk further antagonism from government authorities or litigation, he strategically compromised in a number of key areas, signing off the so-called Kingsbury Commitment of 1913 (drafted by an AT&T Vice President Nathan Kingsbury). This agreement placed limits on how many independents AT&T could acquire, forced the company to divest the 1909 interests it had acquired in Western Union, and stipulated that independents could no longer be denied toll and long-distance services supplied by the Bell system if they requested interconnection. This worked to the Bell system's ultimate advantage because the independents would still pay a fee for using AT&T's lines (Mueller 1997, 129–135). Vail made concerted efforts to avoid the appearance of occupying the whole field of the telephone business, while in reality reasserting *Bell's* monopoly. Independents would be left to develop various areas such as rural markets but would now often be buying Western Electric telephones and technology and be able to interconnect with *Bell*. Vail focused his most vigorous efforts at maintaining more direct control of the most profitable large urban markets and long-distance services. He also accepted a variety of forms of public regulation from state public service commissions who in theory operated to ensure a balance between a fair rate of return and fees charged to telephone users. Gradually the era of two or more telephone companies servicing any given market would pass and by 1915 there were even some legal challenges to value of competition. A judgment of the Supreme Court of Kansas intimated: "Two telephone systems serving the same constituency place a useless burden upon the community, causing sorrow of the heart and vexation of the spirit and are altogether undesirable" (quoted in Winston 1998, 252).

"NETWORK MYSTIQUE" AND TECHNOLOGICAL INNOVATION

During this era, Vail also played an important role in revisiting and further developing the *Bell* systems approach to promoting technological innovation. In particular he encouraged the *Bell* system to not only keep up with emerging telephone technologies by buying up rival patents, which had also been a feature of his earlier term, but also encouraged *Bell* to develop new ideas within its own institutional structures. Vail continued the strategy he had articulated earlier, in 1908, to generate "enough surplus to provide for and make possible any change of plant or equipment made desirable, if not necessary, by the evolution and development of the business" (quoted in Galambos 1992, 3).

Many business historians give Vail credit then, for not only encouraging innovation that was "adaptive," allowing the *Bell* system to be efficiently standardized and responsive to markets, but also, "formative," anticipating and promoting future developments. *Bell's* investments in solving the problems of long-distance telephony would provide an excellent example of "formative innovation." Vail also showed a good understanding of the institutional structures needed to coordinate these different forms of innovation. In pursuing these goals Vail is often given credit for helping develop the idea of "network mystique" and contribute to more general, early twentieth century understandings of the meaning of the term system. Vail described the network and his strategies as, "an ever living organism" [whose development involved] "unceasing effort, continually improving and up building . . . never standing still . . . the plant and methods of each company must be coordinated with those of all of the other companies, because each is but a part of the unified structure . . ." (quoted in Galambos 1992, 4).

Vail's understanding of the role of innovation for the *Bell* system would also encourage the business structure from which the Bell Laboratories would later emerge in 1925. They would become one of the most important sites for scientific and technical innovation of the twentieth century. Vail joined together the AT&T's department of research and development in Boston and the Western Electric's engineering departments in New York and Chicago. While some technical staff stayed in AT&T's central office in New York City, the bulk were merged into a single centralized engineering department housed in New York at Western Electric. This department would later become the Bell Laboratories. Vail appointed John Carty who he had known from his earlier term with *Bell,* as chief engineer. Carty was to help popularize, in somewhat utopian terms, the central role of scientific research for Bell and humanity more generally. He would describe the research laboratory as "a sort of collective mind which, made up of experts in many fields who collaborated continually with one another, could arrive quickly at the solutions of problems so intricate in their ramifications as to require years of single-handed effort, if indeed they could be solved at all single-handed." Carty went so far as to describe the telephone as society's nervous system, " I believe it will be found in any social organism that the degree of development reached by its telephone system will be an important indication of the progress it has made in attaining coordination and solidarity"(quoted in Hoddeson 1981, 530).

THE TRANSCONTINENTAL TELEPHONE LINE

Carty became an important player in assisting Vail to pursue his plan, which emerged in late 1908 and early 1909, to build a transcontinental telephone line. The idea of such a line was advertised as *Bell* fulfilling its longer standing promise that one day the United States would have a unified telephone system with the possibility of calls from coast to coast. Vail authorized the development of the line with the plan that it would be opened at the San Francisco–Panama-Pacific Exposition, which was originally scheduled for 1914 (but ultimately held in 1915). The technology that would be needed to send messages the distance that had been promised was not yet developed. Energy loss and increased distortion appeared in telephone lines as they became longer. By 1893 lines extending 1,200 miles from Boston to Chicago appeared to constitute the limit of what was possible.

During the late 1890s the answer to these problems were provided by the invention of the "loading coil" by George Campbell and Michael Pupin. Loading coils were small electromagnets, that, by being placed at regular intervals along a line, helped maintain the strength of a signal as it traveled along a cable, the details of their optimum sizes and spacing involved a number of important theoretical considerations. Campbell is interesting as a symbol of the emergence of a new generation of telephone inventors with formal scientific training. He studied at Harvard, Gottingen, Vienna, and Paris. *Bell* employed him in 1897, and by 1899 Campbell had developed the theory of the loading coil, taking out his doctorate from Harvard on the topic in 1901 (Hoddeson 1981, 524).

Bell at the opening of the first long-distance line from Chicago to
New York in 1892. Courtesy of the Library of Congress.

The invention of the loading coil would not be immune from the telephone's troubled history of priority disputes over patents. Michael Pupin a
professor of Electromechanics at Columbia University had also been working on the idea of loading coils independently at more or less the same
time (possibly earlier) as Campbell, and took out a patent for the loading
coil in 1900. In 1904 Campbell's and Pupin's claims came before the court.
Campbell was able to provide a more detailed practical and theoretical explanation of their operation, but Pupin convinced the court he had established
the key ideas earlier and won the case. *Bell*, anticipating problems with patent
claims, had already bought the rights to Pupin's patent in 1900, for $185,000
and $15,000 for each year the patent stayed in force (Lubar 1997, 128).

Loading coils allowed lines to be extended, and by 1911 a 2,100-mile line (with loading coils placed at 8-mile intervals) was established between Denver and New York. Loading coils by allowing lines to be constructed by using thinner wire also significantly cut costs. Unloaded lines (lines without loading coils) needed wire approximately 1/8-inch thick, this diameter could now be halved. Prior to 1900, 25 percent of all capital invested in the telephone system had been spent on copper wire (de sola Pool 1977, 28). More work was needed to build a line that would satisfy Vail's promise of a transcontinental line, and *Bell* placed considerable effort into trying to build so-called repeaters, the devices that could amplify a telephone signal. The quest for repeaters and other technological innovations would see *Bell's* research staff grow from 20 members in 1912 to 45 by 1915 (at least 7 with PhDs). In the quest to develop repeaters, *Bell* explicitly sought to systematically draw on the best theoretical science of the day. By 1912, *Bell* made the decision to see if they could adapt a device known as the "audion" invented by Lee de Forest in 1906. The audion consisted of a vacuum tube that *Bell* had inside it a filament which emitted electrons when heated, a metal plate positively charged which attracted electrons, and a negatively charged grid; the latter controlled the flow of electrons between the filament and the plate. Applying a signal to the grid modulated the current and produced an amplified signal in the plate circuit (Hoddeson 1981, 535). Harold Arnold one of AT&T's PhD researchers, would place considerable effort applying new theories of electromagnetism to adapting the audion to the needs of the telephone and helped develop the "high vacuum thermionic tube." Using these repeaters it would finally be possible to build a transcontinental line. *Bell's* research into the repeater also lead to an interest in research into radio technology that became more intense from 1914.

The transcontinental line was 4,300 miles long using four "no. 8" copper wires. It weighed 2,500 tons and was supported by 130,000 telephone poles (Grosvenor and Wesson 1997, 243). Vail put considerable energy into publicizing its triumphant opening. On January 25, 1915, Bell, based in New York, repeated his famous lines to Watson in San Francisco, "Mr Watson, come here, I want you." Watson replied: "It would take me a week to get to you this time" (Grosvenor and Wesson 1997, 246). *Bell Telephone News* hooked into the event as a scientific and also a national American triumph "the highest achievement of practical science up to today—no other nation has produced anything like it, or could any nation. It is sui generis, it is gigantic, and it is "entirely American" (Grosvenor and Wesson 1997, 246).

THE TELEPHONE IN AMERICA AND
THE REST OF THE WORLD

The *Bell Telephone News* could be accused of exaggerating the point that the transcontinental line was an *entirely* American achievement if it is remembered that A. G. Bell was Scottish born and was for much of his life a Canadian resident, but they were correct in a broader sense in that the telephone had emerged and been nurtured in the United States and its refinement and spread as a technological system had outpaced (and would continue to outpace) most other nations until quite recent times with the emergence of the global mobile telephone. It was also true that telephone systems across the world carried an American influence: many countries in the world would establish telephone systems built by *Bell* subsidiaries and derivatives of Western Electric technology. While the telephone was generally adopted in some form very quickly in most other developed nations, its rate of uptake would generally be much slower than in the United States.

A. G. Bell had personally promoted the telephone in Britain demonstrating a telephone to Queen Victoria's approval in 1877. Edison would nevertheless take the honor of taking out the first British telephone patents. A rival firm using *Bell* patents was established and for a brief while there was a fierce rivalry between Edison and Bell-based firms. George Bernard Shaw, who worked for the Bell-based firm, recounted: "Whilst the Edison Telephone Company lasted it crowded the basement of a huge pile of offices on Queen Victoria Street with American artificers. They adored Mr. Edison as the greatest man of all time in every possible department of science, art, and philosophy, and execrated Mr. Graham Bell, the inventor of the rival telephone, as his satanic adversary" (quoted in Winston 1998, 253).

Consistent with the 1879 U.S. settlements between *Bell* and Western Union these two firms would join in 1879 to become the United Telephone Company (UTC). The British Post Office was not happy with the prospect of competition from the possibility of an emerging telephone industry. The GPO took UTC to court and argued against the suggestion made by the UTC that the telephone was a different technology from the telegraph and thus could avoid being covered by the Telegraphy Acts. The court effectively found that the telephone was a form of telegraph. This allowed the GPO to impose a 10 percent royalty on UTC and would control licenses which it would grant to private companies or councils who wished to operate telephone services. The GPO would also be allowed to enter into competition to provide telephone services itself although this would not mean a great deal as they had limited enthusiasm for the development of a system that would compete with their profitable preexisting telegraph service. In 1887 the Postmaster General informed Parliament: "[That] having regard to the cheap and swift means of communication which at present exist by means of telegraph between the principal towns in the U.K. . . . it is extremely doubtful whether there would be much public advantage in establishing telephonic communication generally between those towns" (quoted in Young 1983, 7). The *Times* captured the lack of urgency that accompanied the telephone's spread in Britain when it reported in 1902: "the telephone is not an affair of the million . . . An overwhelming majority of the population do not use it and are not likely to use it at all, except perhaps to the extent of an occasional message from a public station" (quoted in Flichy 1994, 92).

The GPO ultimately took over most of the telephone services and then refused to grant new licenses after December 31, 1911. It then finally took over telephony completely in 1912. In 1914 there were 1.7 telephones per 100 persons in the United Kingdom compared to 9.7 in the United States. In Britain less than 2 percent of the population had telephones. Because of the distractions of World War I, it was not until after 1919 that there would be any significant diffusion of the telephone in Britain (Moore 1989, 232).

In the rest of Europe, with the exception of Scandinavian countries and Germany, who had a slightly higher number of subscribers than Britain, the spread of the telephones during this era was slower yet again. While the spread of the telephone in Britain paled behind the United States, in most of Europe the rate of diffusion of the telephone was even lower. In 1906, Britain with a population of 42 million had more working telephones than the combined 288 million people spread across Austria, Hungary, Belgium, Denmark, Holland, Italy, Norway, Portugal, Russia, Sweden, and Switzerland. Most of Britain's telephones were in the large cities such as metropolitan London (Moore 1989, 232).

Adding to factors such as government resistance to the telephone, and support for the better entrenched monopoly telegraph systems, the situation in Europe has also sometimes been explained in terms of a lack of demand for telephones because of the persistence of traditional rural and village modes of life. For example, many parts of Europe offered a different pattern of population than that emerging in the United States. Whereas, the United States had a combination of large cities, emerging suburban developments, and in many places a widely geographically dispersed rural population, a feature of many parts of Europe were towns and villages with relatively small but densely settled population and farms operating in relatively close proximity to one another. Traditional forms of communication were under less challenge than in the more rapidly "modernizing" environment of the United States (Flichy 1995, 92–93).

The spread of telephones in 1914 across the world shows a similar pattern to the ratios of phones in the United States relative to Europe and Britain. For instance there were 2.8 telephones per 100 persons in Australia, 3.5 in Hawaii, 4.6 in New Zealand, 6.5 in Canada, and 9.7 in the United States (Young 1983, 23).

THE BUSINESS TELEPHONE AND THE "PROBLEM" OF UNRULY USERS

While the telephone in this era was becoming an increasingly familiar technology in the United States in particular, and albeit more slowly in most other developed countries, the question of the best ways the telephone should be put to use was still being negotiated by telephone promoters and users. In general, despite the rhetoric of "universal service" most promoters of the telephone up until the 1920s still thought of the telephone in the tradition of the telegraph as first and foremost a business tool that could also transmit orders, messages, and be useful in emergencies. When sales were less than expected they routinely blamed users for not being sufficiently educated in the uses of the telephone. Users of the telephone had begun to apply the telephone to a much wider variety of communication purposes than promoters envisaged. The most fundamental one, which today seems obvious, was extended private conversation, or "sociability." The match between the expectations of users and promoters would gradually come together in the 1920s.

Early promotional campaigns and telephone trade magazines emphasized the variety of practical functional services that the telephone could offer. These included weather reports, sports results, firefighter alerts, and baby lullabies. Formal advertising that commenced around 1910 was geared toward businessmen and emphasized the role of the telephone in saving time, planning, impressing customers, being modern, and keeping in touch with work while on vacation. One widely cited advertisement from 1914 and 1915 suggested "[y]ou fisherman who feel these warm days of Spring luring you to your favorite stream. . . . You can adjust affairs before leaving, ascertain the condition of streams, secure accommodations, and always be in touch with business and home" (Fischer 1988, 40).

A second major theme which appeared in promotion of the telephone in this era was "household management and planning." Advertisements suggested subscribers could keep in contact with work, issue invitations, and send messages or orders to schools, dressmakers, coal vendors, and plumbers (Fischer 1988, 39). The emphasis on practical home management and business uses also encouraged *Bell* and other phone companies to phase out party lines. In the early years of telephone development it was not uncommon for a subscriber to share a line with a number of other users that passed through an exchange. The way these lines were grouped together could be quite arbitrary. Businessmen increasingly asserted their need for privacy. Before the technology could gradually be reconfigured to reduce the number of party lines various rules of conduct relating to eavesdropping and appropriate manners for conducting phone conversations were set by *Bell* and even given some legal support: although how easy such codes of conduct were to "police" in practice may have remained difficult (Fischer 1992, 70–71).

Users of telephones did not always comfortably conform to the industry expectations; women users and rural users of the telephone in particular did not behave in the way they were meant to. Wives of middle-class and upper-class businessmen began using telephones extensively for "day-to-day" conversation and sociability, not just for messaging and sending orders, and telephone conversation was becoming a cultural activity in its own right. In some residences supplementary lines and extensions were gradually being built to accommodate the telephone's dual use as a business tool and as tool to enhance sociability (Martin 1991, 318–320). Some telephone promoters initially responded disparagingly to such supposedly frivolous uses of the telephone. After listening to a sample of calls from the exchange, a local telephone manager in Seattle in 1909 determined that 30 percent of calls were " purely idle gossip," 20 percent orders to stores and business, 20 percent from subscriber homes to their own businesses, and 15 percent social invitations. The manager believed that these types of ratios were representative of other cities and exchanges. The high percentage of gossip calls was defined as an "unnecessary use" and something that needed to be eradicated by education programs (Fischer 1988, 48).

Rural users also resisted the expectations of telephones promoters. Because of the greater influence of independents and lower investment of infrastructure on the part of Bell, single party lines were common in rural areas. Isolated rural women in particular began using the telephone for activities such as "meeting on the lines" and some contemporary commentators began to note the value of the telephone in helping to diminish the feelings of isolation of rural women. Party lines in rural areas also involved users forming their own patterns of eavesdropping practices and negotiating their own notions of its acceptability. In 1907, a newspaper in North

Dakota described the eavesdropping culture of the party line in the following terms: "Usually when a country subscriber rings anyone up several of his neighbors immediately butt in—not to talk—just listen. . . . Then there are a number of persons gossiping by the way of the telephone, and the business of T. Roosevelt, even, would have to wait, once they get started, till the matters of the entire community have been wafted over the wires. And occasionally a real talk-fest occurs when there isn't much difference in the cyclone of conversation and the flow of a sewing circle" (quoted in Kline 2003, 54).

Many of the disparaging commentaries were framed in gendered terms emphasizing that the practice seemed to be dominated by women. A 1914 issue of the *Literary Digest* published a photograph of a women with a telephone receiver tied to her head by a piece of cloth so that she wouldn't miss a word while she sat at her sewing machine (Kline 2003, 55). Some feminist commentators have noted that the highlighting eavesdropping as a female habit reflected stereotypes about the "natural" tendencies of women but more importantly reflected the reality of the important role that women play in holding informal community networks together (Rakow 1988). Recalling the attitudes of that time, one farming woman described the typical conduct of a party line: "A lot of times when you were in a conversation, somebody would come on the line and say, 'Is that you Mabel/ Do you know your cows are out?' Or, 'Are you going to be home?' or something like that. Pretty soon you'd have three parties on the line or sometimes four" (quoted in Kline 2003, 56).

Other reports from the time suggest that men were also not averse to eavesdropping. Many managers of local systems complained among other things that eavesdropping tied up lines limiting the volume of possible calls that could be made between intended senders and receivers and wore out telephone batteries. They replied by promoting various methods to try to stop eaves dropping. Strategies included: trying to extract extra call charges by monitoring telephone battery use, fining eavesdroppers, limiting calls to 5 minutes, giving business calls priority. States such as Ohio and Indiana even passed laws which made repeating the contents of an eavesdropped telephone conversation a crime and telephone industry journals printed cartoons, poems, and newspaper stories recounting the harm caused by eavesdropping (Kline 2003, 55).

Some telephone engineers attempted to develop technologies to overcome eavesdropping. Many of these devices were nevertheless expensive and not particularly practical. There were also some cases of engineers, in a sense conceding defeat, who upgraded the standard induction coils in some farmers' lines to help increase the audible volume of calls under the

From the 1920s and onwards telephone companies began to promote the potential of the telephone to promote "sociability." "Wife of ranch owner on a party line 1930s." Courtesy of the Library of Congress.

assumption that there would be multiple listeners (Kline 2003, 58). From the 1920s onward, promoters of the telephone would increasingly synchronize their promotional campaigns to note the telephone's potential as a tool to enhance sociability, something revealed to them by these "unruly" women and rural telephone users.

5

Consolidation in the Interwar Period, 1918–1945

The pattern established by Vail and the Kingsbury Commitment was to prove a robust model for the telephone business and despite some variations and challenges its influence would still be felt into the 1980s. The First World War offered a brief interruption to the Vail model, as *Bell* was briefly nationalized in 1918. After the war, with calls for cheaper services, the Kingsbury Commitment returned as the template for the organization of telephones. After Vail's departure from Bell in 1919, in 1921, the Willis-Graham Act cemented the rationale of the Kingsbury agreement in law allowing *Bell* to be exempt from antitrust limits on purchasing telephone companies (Mueller 1997, 145). *Bell* continued with the aims set by Vail of continuously spreading service and responding to the government's lead with pricing. During the depression for instance, *Bell* cut its fees by 5 percent at the government's request. The consolidation of the Kingsbury Commitment by the Willis-Graham Act reinforced Vail's vision. *Bell* tacitly recognized that maintaining its profit margins and its protected position would be politically dependent on its continued investments back into the system. This helps explain its continued investment in technology and the emergence of the Bell Laboratories in 1925. The Bell Laboratories were both an economic and a political investment. Following the Great Depression, Roosevelt's New Deal Government took greater interest in regulating the telephone system. Public Utility Commissions, who could

Bell laboratories officially came into being in 1925 and would become one of the most important sources for technological innovations in communications of the twentieth century. "Inside Bell Laboratories." Courtesy of the Library of Congress.

evaluate service, and pricing appeared in every state, and finally, in 1934 the Federal Communications Commission (FCC) was established by the Communications Act (Faulhaber 1987, 7).

During this era *Bell* was generally viewed favorably and accepted by many as a natural monopoly. However, it was still never free of critics: some wanted competition, and others, who looked at the dominant patterns for organizing telephones outside of the United States, suggested some form of nationalization would be better than *Bell's* corporate monopoly. The Walker Report of 1938, for example, described the pattern of regulation surrounding the *Bell* system as unworkable and suggested that some form of nationalization would be preferable. The report had criticized *Bell's* vertical structure and, in particular, Western Electric's sheltered position as a supplier of components to the telephone system. At the time the Walker Report received a "luke-warm" reception, but similar critiques of *Bell* (particularly its relationship with Western Electric) would reappear over the next 2 decades (Faulhaber 1987, 7–8).

THE EMERGENCE OF RADIO TECHNOLOGY

The emergence of radio technology at this time would also constitute both challenges and opportunities for *Bell*. From its earliest days radio technology had been of interest to Bell Laboratories although this would initially be pursued rather quietly until 1914. Preoccupied with maintaining its protected-monopoly telephone position, *Bell* may have been concerned not to appear to be moving too quickly into yet another area of communication technology. Given *Bell's* success, it was also unlikely that radio would have been seen as a competitor (Hoddeson 1981, 538–541).

As was noted in the last chapter, the core technologies that had made the transcontinental telephone line a possibility were the audion and vacuum tube. Developments in vacuum tubes were also central to the creation of radio technology. Many of these early developments in radio technology involved patents held by *Bell* and a number of rival companies. The relevant parties realized that they had come to something of an impasse and would need to do something if they were to be able to develop radio technology. This awareness led in 1920 to the major players signing a cross-licensing agreement. AT&T, General Electric, and the Radio Corporation of America, were the original signatories covering 1,200 patents (Westinghouse also entered the agreement in 1921). Each party agreed to grant rights to the other to use patents but limited the markets to which each party could apply their technology. *Bell* agreed not to enter into the actual business of radio broadcasting but, in return, would be able to maintain exclusive control over public markets for radiotelephony and its existing wires.

This cross-licensing agreement also meant that *Bell* could devote major energies into improving and mass-producing telephone technologies, devote research into better forms of switching, and transmission capacity via Bell Laboratories, and also buy up or trade patents with other firms who were not in a position to compete in the telephone business. These arrangements added to *Bell's* already lucrative financial base and, in 1929, it became the first company in the United States to generate $1 billion in revenues.

This era would also witness the steady decline of manually operated switches although it is important to remember they did not vanish "overnight." Even with dial telephones and Strowger switches, many operators still provided forms of directory assistance. Western Electric had bought rights to Strowger switching technology in 1916 but would only start developing automatic exchanges in earnest from the 1920s. The ever-increasing volume of calls made to large city exchanges made manual switching increasingly untenable, and with customers becaming increasingly familiar with using the dial telephone, notions like personal service were no longer a marketing point, as users favored the greater speed and privacy of going through an automated exchange. In 1938, the first crossbar switching systems also offered further improvements to the common Strowger-based switching systems (Farley 2006).

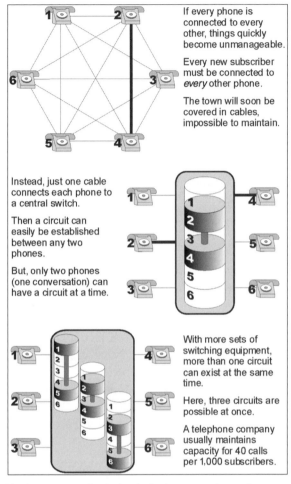

The basic principles behind telephone switching. Courtesy Robert B.K. Brown, 2006.

During this time, radiotelephone services also expanded. *Bell* was able to further cut costs for longer distance services and by 1930 had halved the cost of a call that had previously only been available by land line between San Francisco and New York: still relatively expensive at $10 per 3-minute call (Lubar 1993, 134). Another important technological improvement was the development and use of coaxial cable from the 1940s. These cables offered much better insulation allowing a greater range of frequencies to be transmitted and in turn a much greater quantity of information to be carried: they would be of particular value for improving longer line service and television.

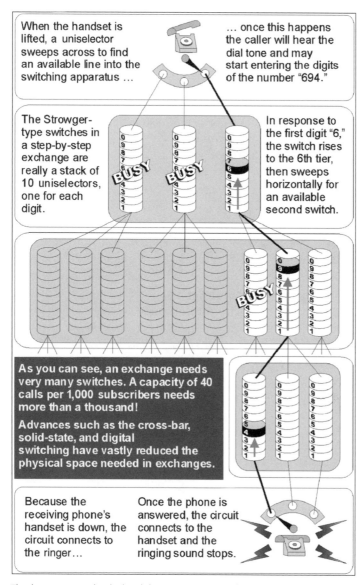

When the handset is lifted, a uniselector sweeps across to find an available line into the switching apparatus ...

... once this happens the caller will hear the dial tone and may start entering the digits of the number "694."

The Strowger-type switches in a step-by-step exchange are really a stack of 10 uniselectors, one for each digit.

In response to the first digit "6," the switch rises to the 6th tier, then sweeps horizontally for an available second switch.

As you can see, an exchange needs very many switches. A capacity of 40 calls per 1,000 subscribers needs more than a thousand!

Advances such as the cross-bar, solid-state, and digital switching have vastly reduced the physical space needed in exchanges.

Because the receiving phone's handset is down, the circuit connects to the ringer...

Once the phone is answered, the circuit connects to the handset and the ringing sound stops.

The basic principles behind the Strowger Switch. Courtesy Robert B.K. Brown, 2006.

HENRY DREYFUS, INDUSTRIAL DESIGN, AND THE BELL MODEL "300"

Bell's financial stability and effort for standardization of equipment also witnessed the consolidation of a number of core simple telephone designs such as those produced by Henry Dreyfus and Associates: one of the most famous would be the rotary dial "300" and later "500" telephone series. These designs were noted for their simplicity and elegance and became one of the most immediately recognizable technologies of the twentieth century. Some have made the comparison with the model "300" as the template for the "ideal" telephone and the model T Ford as a template for the "ideal" automobile. Henry Dreyfus (1904–1972) was born in New York City. He is often considered alongside Raymond Loewy and Walter Dorwin Teague as one the pioneers of industrial design in the United States. Some have argued that interest in industrial design had been stimulated in the recession and depression where companies were desperate to distinguish their products from their competitors. It also hooked into a market that was becoming more sensitive to the idea of products appearing modern. Some of the key industrial designers, such as Dreyfus, differed from traditional engineers and designers in that they had originally developed skills in artistic and commercial settings. Dreyfus, for example, had started out as a successful Broadway stage designer before setting up his 1929 industrial design office.

In 1927 *Bell* had introduced the handset telephone but was eager to come up with the best standard design for the telephone as soon as possible. *Bell* wanted a telephone that would be functional for the user, but it also wanted a design that would appear modern, be durable, and allow easy repair. In 1930 *Bell* offered $1,000 to 10 artists to develop ideas for future telephones. Dreyfus was approached by *Bell* but initially declined to engage in the exercise: "I suggested that a telephone's appearance should be developed from the inside out, not created as a mold . . . and that this would require collaboration with *Bell* technicians. My visitor disagreed, saying such collaboration would only limit a designer's artistic scope" (quoted in Weed 1996). *Bell* feared that Dreyfus would become artistically compromised if he worked too closely with engineers. The other designs had proved impractical. So after some months, *Bell* again approached Dreyfus who this time got his way and collaborated with *Bell's* engineers, manufacturers, and even their repairmen, "[b]ecause placement had a bearing on design, we had to determine what people did with phones, and that's why the telephone company permitted me to act as a repairman's helper when he went on his rounds" Dreyfus also recalled "some people weren't quite sure where to put [telephones] . . . [t]hey were sometimes kept inside plaster globes of the world or cabinets or dolls with fluffy skirts" (Stern and Gwathmey 1994, 41).

Dreyfus tested the prototypes for his telephone design with *Bell* engineers and dozens of users, making measurements of people and telephones. His approach was one of the earliest examples of what later became known

1928 Desk Set

1937 Dreyfus 300

1983 Touch-a-Matic 1600

Different generations of the modern telephone. Courtesy
Robert B.K. Brown, 2006.

as the science of ergonomics. Some of this work would appear in his book
The Measure of Man. The result of Dreyfus's labors was the Bell "300" in
1937 (and after WWII, in 1949, the Bell "500"). The "300" had 10 finger
holes in a rotary dial, with numbers in black and letters in red. As the dial
turned it would gently click. The bell was enclosed in the square base of the
phone, the body was solidly molded, and the standard color was black. The
basic design would be modified in subtle ways and different materials would
be used as time went on especially as plastics became increasingly cheaper
and easier to mold, but they became the iconic design for telephones during
the better part of the twentieth century (Weed 1996).

PROMOTING THE SOCIABLE TELEPHONE, THE 1920s AND BEYOND

It would be in this period that the telephone would become a much more standard technology for many people in the United States. By 1929, 42 percent of all U.S. households had telephones: a figure that declined during the Depression to 31 percent and reconsolidated to 37 percent by 1940 (Fischer 1988, 36). Despite these generally steady trends of ongoing diffusion of the telephone, it is important to note the decline in telephone use in rural areas from a time immediately prior to the Depression (which did not pick up again till after World War II). This decline was greater than in urban areas and appeared to be a response to more than just the general economic woes of the time. Rural users who had been one of the most enthusiastic adopters of the telephone seemed to be equally willing to, for more than a decade, reject it. A number of explanations that go beyond simple economic hardship have been put forward for why this short but interesting run against the general pattern of telephone history occurred. During the period of growth in telephones in rural areas, farmers also began to adopt other important technologies such as basic electrical appliances and most importantly the automobile. Some of the functions of the telephone, such as overcoming rural isolation, may have, because of these other technologies, been felt less acutely. These other technologies in a sense acted as substitutes for some of the functions previously satisfied by the telephone. Another explanation links declines in telephone use with technical upgrading and standardization of telephones in rural areas. Much of the first generation of rural telephones had relied on smaller co-ops which were often cheap but used poorer technology and offered lower quality service. Upgrading and linking rural telephone lines would ultimately improve the quality of service and allow rural areas to more readily be part of the broader telephone network, but this would lead in the short term to more expensive telephone charges.

Aside from these interruptions to more general patterns of change, it was in this era that the telephone would increasingly be promoted by *Bell* as a device that would enhance sociability and not just act as a business tool. As noted in the last chapter during the earlier period of the emergence of the Bell system, telephone promoters had been reluctant to acknowledge that many users of the telephone were embracing its potential to enhance sociability and that this was a legitimate way for the telephone to develop. Promoters of the telephone from the 1920s started advertising the telephone focusing on the positive subjective and emotional consequences for the user. In a trade manual in 1923 the Southwestern Bell company announced that it was selling something: ". . . more vital than distance, speed and accuracy . . . [T]he telephone . . . almost brings [people] face to face. It is the next best thing to personal contact. So the fundamental purpose of the current advertising is to sell the subscribers their voices their true worth—to help them realize that "Your voice is You." . . . to make subscribers think of the telephone whenever they think of distant friends or relatives. . . ." (quoted in Fischer 1988, 41).

Sociability started to appear as a theme to promote long-distance calls but it would gradually spread to the promotion of the comfort and convenience of the telephone for "day-to-day" uses. *Bell* may have observed that by the late 1920s, American families had embraced the automobile, electrical appliances, and gas services more fully than the telephone, and it began to borrow from some of the marketing strategies of the automobile. After the interruptions of the Depression years these themes would be revisited with increased energy. In 1935 an advertisement posed the questions, "Have you ever watched a person telephoning to a friend? Have you noticed how readily the lips part into smiles . . .?" In 1937 another suggested "Friendship's path often follows the trail of the telephone wire" and in 1939, "Someone thinks of someone, reaches for the telephone, and all is well" (quoted in Fischer 1988, 43).

Guides for telephone salesmen noted the telephones importance in the case of emergencies but also increasingly emphasized the telephones "sociability" potentials. Similar guides in the earlier era (1904) made no mention of conversation and sociability. A 1931 sales guide advised under the heading of "Fosters Friendships" . . . "Your telephone will keep your personal friendships alive and active. Real friendships are too rare and valuable to be broken when you or your friends move out of town. Correspondence will help for a time, but friendships do not flourish for long on letters alone. When you can't visit in person, telephone periodically. Telephone calls will keep up the whole intimacy remarkably well" (quoted in Fischer 1988, 45).

The adoption of the telephone as a convenience and gradually a necessity, along with the automobile, became a pattern broadly accepted by promoters of the telephone and its users, and would be a feature of "the life" of the telephone in the decades between WWII and the divestiture of *Bell* in the early 1980s. By the eve of the Second World War, *Bell* controlled 83 percent of all U.S. telephones, 98 percent of all long-distance wires and was the world's largest firm with $5 billion assets (Winston 1998, 249).

6

The Calm before the Storm, 1945–1970s

◆

The post–World War II period till the late 1960s and early 1970s represented an era of ongoing stability for the users of the telephone and its basic meaning and functions and, for this reason, the telephone during this time can be thought of as "the standard" telephone. This was also an era where a series of important new communication technologies would emerge. These key technologies would be the transistor; information theory; digital information transmission; satellites; and computers. There is also an interesting example of a technology that designers and promoters invested huge energy in and which was never adopted, "the Picturephone." Bell Laboratories was one of the key sites where many of the ideas for these new technologies emerged. Despite their radical potentials most of these innovations would *initially* have a relatively small impact on the telephone and telephone users: satellites, and microwaves, provided improved paths of transmission and there were significant, incremental improvements in cost and quality of service through things like digital switching technologies and improved handsets. In some respects, for the telephone, this period marked the "calm before the storm" that would be created by the deregulation of telecommunications and the take off of microelectronics of the late 1970s and early 1980s.

THE TRANSISTOR

Developing more efficient switching devices and better systems for amplifying signals had been a long-term project of the Bell Laboratories. Improving these technologies were seen as things that could foster continued improvement in traditional telephone services and also radiotelephony. The war effort had also provided a considerable incentive to develop these technologies as quickly as possible. Out of this context, on July 1, 1948, Bell Laboratories released one of the most important technologies of the twentieth century: the transistor. This device was the joint invention of William Shockley, John Bardeen, and Walter Brattain. The transistor was the forerunner to the microchip which has allowed the continuous miniaturization and increase in the power of computers.

Transistors (from the idea of transit-resister) function as miniature switches that provide a much more resilient, stable, and compact alternative to vacuum tubes/valves. The transistor, in the simplest terms, operates by controlling the amount of current that can flow between two terminals by a voltage applied to a third terminal. A stronger signal (for example 10 watts) can be directed into one side of the transistor that is stopped (resisted) by a poor conducting material such as silicon. A weaker signal (for example 1 watt) can then be directed into the middle terminal. Because the silicon has had various chemical impurities strategically introduced into it, the weaker signal triggers the silicon to start behaving as if it is a conductor and it allows the strong signal to pass (transit) through it. As this stronger signal passes through, it now also carries with it the weaker signal. Depending on the way particular impurities are added to semiconducting materials, such as silicon, various kinds of transistors with different switching and amplifying qualities can be constructed. Using different transistors, electrical circuits can be put together far more compactly and with greater reliability and durability than traditional switches and vacuum tubes (Farley 2006).

Apart from transistors being added to technologies directly involved with the telephone itself, they would constitute one of the core technological innovations that would make the later boom of the 1970s and 1980s in electronic industries possible. It took a number of years for transistors to be directly applied to the telephone: but from the outset many commentators were excited by their potential. One of the most important applications of the transistor to the telephone system was to offer ways to assist in the construction of more reliable switches that could handle much larger volumes of calls. One key system was the so-called stored program telephone-switch system. It was first put into commercial use in 1965 after approximately 30 years development and $500 million of investment The system operated using more than 90 million input/output stations and its developers spoke of each system as a form of computer (Lubar 1993, 135).

MICROWAVES

Microwave technology had been given a considerable boost during the war in the race to develop better radio communication systems, and in particular radar. *Bell* had been a significant contributor to the war effort as had been a number of other smaller firms such as Philco and Raytheon. Apart from long-distance telephony, one of the main commercial applications of microwave technology would be its use to facilitate the rapid expansion of television services. The cross-licensing agreements of the 1920s, which divided up radio broadcasting from radiotelephony, and which allocated *Bell* the monopoly on providing infrastructure, would be used to stifle the claims of these smaller firms to provide microwave infrastructure to carry expanding television services. *Bell* worked hard to keep competitors out, arguing to the FCC that they should be the preferred choice to provide microwave and cable services for broadcasting, as a share of the profits from these lucrative new services could be used to subsidize lower rates for telephone users and thus support *Bell's* ongoing "project" of providing universal service (Faulhaber 1987, 25).

INFORMATION THEORY

Information theory offered a set of conceptual tools that assisted the development of digital computers. One of the key players who helped information theory to take on this role was Claude Shannon of the Bell Laboratories. In 1948 he wrote the Mathematical Theory of Communication. While mathematicians had for years explored ways of encoding information, Shannon's work inspired a growth in interest in the development of sophisticated models to measure information. A major preoccupation of these models was to establish the most efficient ways a message could be sent along a channel with as little distortion as possible. From this work it was realized that there were a number of ways that the "information content" of an original message at the sender's end could be radically compressed and minimized and still be reconstructed meaningfully at the receiving end of the channel. Developing increasingly better ways of coding and decoding information, such as turning information into a digital form (ones and zeros) meant the quality of the transmission of a signal was far less crucial than in traditional analog systems, which converted an existing message into an electronic signal but didn't compress it first. Digital data transmission and the development of the Internet would benefit from this work. It took a number of years, till 1956, for *Bell* to first transmit digitally and a number of further years were needed to iron various technical problems (Lubar 1993, 158).

FIBREOPTICS

Corning Glass pioneered the development of fibreoptic technology in the early 1970s. In this system, information was carried by modulated light through glass cables rather than by electrons traveling along copper cables. It is interesting to note that carrying information via light had been one of A. G. Bell's research interests after he moved his interest away from the traditional telephone. Coupled with lasers which were able to modulate light at extraordinary rates and produce digital (on/off coded) signals, fibre optic cables would, after a decade of heavy investment by *Bell* and others, by the 1980s challenge traditional forms of wiring and radio broadcasting for the speed and volume of information they could carry (Flichy 1995, 134–136).

SATELLITES

Yet another important technology that would expand *Bell's* post–Second World War telephone capabilities was the development of satellites. The idea of satellites had been expounded by Arthur C. Clarke in 1948. The less well-known *Bell* scientist John Pierce, a number of years later, would help to actually put the ideas into practice. In 1962, the *Telstar* satellite designed by the Bell Laboratories would be launched. By the mid-1970s there were numerous satellites in operation: *Comstar*, launched in 1976, could carry 30,000 calls at the same time. Between 1974 and 1975 alone, the number of transatlantic phone calls increased tenfold (Lubar 1993, 137).

THE PICTURE PHONE, A PATH NOT TAKEN

The broad image that emerges from surveying *Bell* in the 3 decades following World War II is one of continuing successful developments in new technology, developments that would not only expand the capacity of telephone network and improve service but also contribute to the emergence of the development of information technology more generally. While this image is generally correct, a reminder that the process of technological change is not a simple story of linear advances can be found with the invention, attempted development, and commercial failure of the picturephone. At the New York World Fair of 1964, *Bell* displayed a model of the picturephone. The picturephone was basically a device that allowed callers to see each other on a small TV-like screen while they conversed. While the general idea of transmitting visual images along the telephone line had been possible for more than 20 years, *Bell* now believed the technology was sufficiently mature to be developed. Their beliefs in its commercial viability were boosted when in a survey of visitors to the World Fair, 60 percent suggested that it was important or very important to see the person they were talking to. Despite *Bell* investing over $500 million in commercial development of the picturephone, the technology turned out to be a financial disaster: practically none were sold. It would appear that when it came to the reality of actually using the technology, seeing a person on the telephone while talking to them simply was not considered worth the extra cost and it did not add enough to the amount of relevant information being transmitted to be worthwhile (Lubar 1993, 134).

DIFFERENT USES OF THE STANDARD TELEPHONE

The emergence of these important new technologies would ultimately have a significant impact on the place the telephone would occupy in modern telecommunication infrastructures, but they would not exert a significant *immediate* impact on the way telephones were used or thought of. The basic idea of the function and form of the telephone would become more or less stable as it became ubiquitous across post-World War II America and other developed countries. In a period of postwar economic growth, and with the aid of government subsidies to help return telephones to rural areas, *Bell* experienced difficulty in keeping up with the demand for telephones. In 1950, 62 percent of residences held telephones subscriptions, 80 percent by 1962, and 90 percent by 1970 (Fischer 1992, 53).

Putting forward broad generalizations about the social impacts of the telephone even during this long period of stability of form and function is possible but something that must be done with some caution. It is important to avoid speaking of the impacts of technology as if technology is

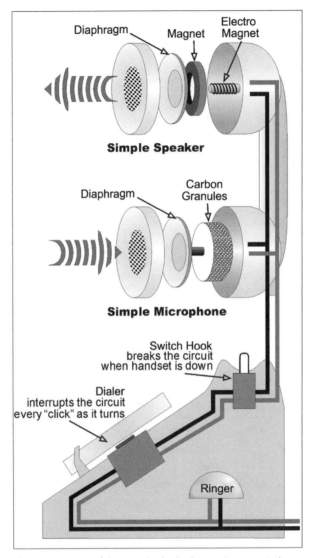

The components of the standard telephone. Courtesy Robert B.K. Brown, 2006.

somehow independent of society and that "impacts" can be simply "read off" the "logic" of the "physical attributes" of any given technology. As discussion in past chapters has shown, users may respond to and adapt technologies in different ways, and in ways different to the way that designers and promoters of technologies anticipate (Oudeshorn and Pinch 2003). For example, women and many rural users of the early telephone adapted what was intended to operate primarily as a business tool, to a device for extended conversation and sociability. These informal "negotiations" between designers, promoters, and users form an important part of the history of the telephone, and are not unique to patterns observed in the development of technologies more generally, although the significance of the user in helping shape the telephone may be something that is generally more typical of consumer technologies (de Sola Pool 1983, 14). Social pressures encouraging different patterns of use in different societies of the same basic technology also provide a cautionary note to simply "reading off" social impacts from the "physical attributes" of a technology. A brief history of the role played by the telephone in the Amish society provides an instructive example.

THE AMISH AND THE COMMUNITY TELEPHONE

The Amish first settled in Pennsylvania in 1737 fleeing from religious intolerance in Europe. They established communities based on farming, oral traditions, strong community, and religious observance. They have become well known through their reluctance to embrace modern technology or fashions that they believe will destabilize their community traditions. When the telephone became available in the early twentieth century, the Amish were initially indifferent to considering its implications. In 1910, it nevertheless came to be one of the subjects of a major split in their community. Some Amish valued the convenience of the telephone, yet others believed that it was a threat to community values. In particular the telephone allowed the entry of the outside world into the homes of the close knit communities, exposed individual community members to business with the outside world, and some feared it would break down their strong traditions of oral face-to-face communication. There were also particular concerns with the possible negative impacts on women and youth. Apparently feelings ran strongly enough for the dispute to encourage some members of the communities to leave. In the wake of these disputes the Amish banned the telephone for decades.

In the mid-1930s they revisited the question of whether or not they should allow telephones. There was something of a consensus that at least some access to the telephone was useful for emergencies. From these negotiations resulted a compromise that lead to a distinctively "Amish pattern" of telephone use. The Amish developed the idea of the community telephone or "telephone shanties." Telephones would only be located in small sheds separate from the house, out of the way, near the end of lanes, or beside barns. They had unlisted numbers, would be used almost exclusively for outgoing calls (loud bells indicating incoming calls would be banned), and they would be shared between half a dozen families or more from any given neighborhood (Zimmerman Umble 1992, 184).

SOCIAL "IMPACTS" OF THE STANDARD TELEPHONE

The difficulties in producing simple cause and effect accounts of the social impact of technologies need to be kept in mind when making generalizations about the impacts of the telephone during its long period of stability of basic form. Three related sets of questions tend to reappear in much of the relevant literature:

First, what influence did the telephone have on social space? Did it weaken a sense of local community?

Second, did the telephone expand or diminish social relations?

Third, did the telephone have identifiable psychological effects on users, in particular, in relation to perceptions of privacy and security and anxiety? (Fischer 1992).

TIME, SPACE, AND LOCAL COMMUNITY

It has almost become a cliché in some texts to note that from the first half of the twentieth century one of the consequences of "modernity" was that local cultures in developed countries, the United States in particular, became steadily absorbed into being part of a mass culture. People experienced fewer attachments to their locality and became more cosmopolitan and less parochial in their outlook. The telephone (along with the automobile, the radio, and later, the television) contributed to this process by allowing people to widen their circle of contacts and not be restricted by the traditional limitations of time and space. A positive view of these "delocalizing" processes suggest that there were opportunities for social ties to be widened and enriched, local prejudices to diminish, local accents/dialects to decline, and broader political vision to develop. In some respects these views reflect a less extreme version of some of the utopian claims that always seem to have adhered to assessment of the social impacts of communication technologies (Mcluhan 1964, 233–240). From a negative vantage point, these "delocalizing" processes allowed communities to be broadened, but at the cost of them becoming shallower and more artificial, with people more easily able to disassociate themselves from real concerns in their local neigh-borhoods. Some have challenged this "delocalization" thesis altogether, both negative and positive versions, and suggested that the telephone as a "point-to-point" form of media that allows feedback and spontaneous en-gagement encourages people to actually strengthen their local ties. This role can be contrasted with other media such as radio and movies that encourage a sense of passivity and placelessness.

Plenty of advertisements and anecdotes can be assembled to back up all these sets of propositions. Detailed long-term sociological surveys, nev-ertheless, have suggested more mundane conclusions (Fischer 1992). They suggest that the telephone in U.S. culture did appear to facilitate people making wider contacts more easily, and more frequently engage in "extra-local" activities, but most of these activities didn't appear to correspond with any radical alterations to lifestyle. People gradually made more long-distance calls, but this wasn't done at the expense of using the telephone to maintain local ties. Overall, the widespread use of the telephone *did* appear to be part of a subtle increase in social activity more generally, and fitted in with broader trends of greater concern with the outside world and greater cultivation of the private sphere of the household, but it's widespread use *did not* correspond with a marked decline in concern with local matters (Fischer 1992, 220–221).

DEEPER OR SHALLOWER SOCIAL RELATIONS?

As was noted in Chapter 5, from as early as the 1920s and the 1930s, telephone advertisements emphasized the qualities of the telephone as a device to enhance and enrich personal relationships (Fischer 1988). In this tradition the telephone's capacity to overcome distance can be seen as a vital way that people have been able to sustain social relationships that otherwise may have been difficult or impossible to maintain. Feminist historians of the telephone tend to have supported this positive image by focusing on the special role that the telephone has played for women (Rakow 1988). A variety of surveys by industry and social scientists across different nations have supported the thesis that women have had a special affinity with the telephone. Women dial the greatest number of long-distance calls, spend longer on calls more generally, and are more likely to call family and friends than men (Fischer 1992, 231). Interviews have also reinforced that telephone users themselves often identify the telephone as forming a more important part of the domain of women than men (Moyal 1995, 284–310).

A number of reasons have been put forward to help account for the affinity of women and the telephone. First, for much of the period of the consolidation and stabilization of the dominant form of the telephone, women have been more likely than men to have been engaged in the domestic sphere, for instance, performing child care from the home. Satisfying such roles often lead to significant periods where contact with the outside world could be difficult. The telephone was able to constitute something of a "life raft" enabling social links outside the home. Second, women have been more likely to be cast in the role of being "social managers," either, in the traditional role as home manager, or in workplace settings as administrators, and secretaries. In satisfying these roles they have been more likely to have been given the tasks of organizing meetings and maintaining family contacts: activities encouraging greater use of the telephone. Finally, women have been recognized to generally be more comfortable and adept than men with verbal emotional styles of communication, attributes compatible with sociable use of the telephone.

Some of the detractors of the telephone's influence on social relations have suggested, aside from any kind of affinity between women and the telephone, that on balance it has had subtle negative effects on community and sociability. This negative thesis emphasizes the fact that people have steadily replaced personal contact with telephone calls. Contact can be made across wider spaces and more frequently but they are less personal and increasingly shallow. Dispersed shallow sociability also leads to the possibility of the devaluation of the physical public spaces which were important for traditional "face-to-face" forms of communication. Rather than visit friends and family in person, a quick "emotionally superficial" telephone call can be made. Families have less incentive to live in physical proximity because they can still keep in some form of contact across growing suburban sprawls. Similar factors also apply in rural settings, farmers with telephones now have less reason to travel into a town to communicate. In turn, small towns and formal meeting places have become run-down and less significant. So rather than offer the solution to the problems of rural isolation (as has often been thought) the telephone may have actually ended up contributing to them. The plausibility of these negative assessments hinge, in part, on whether or not telephone contacts are seen as constituting substitutes to "face-to-face" communication, augmenting existing forms of communication, or offering new possibilities for communication. For instance, if telephones were unavailable it might mean that certain contacts between people simply would not occur (de Sola Pool 1983, 129–130).

There are obviously some difficulties of assessing in any definitive sense, these claims about the impact of the telephone on the depth of social relations. By their nature social relations are difficult to measure and it is difficult to make assessments without engaging in subjective judgments. What is more measurable is that users of telephones are more likely to be making calls for social than practical reasons. Research done by AT&T has suggested that most telephone use is restricted to a small circle of friends or family, on average to only five numbers. Surveys done during the 1980s suggested three-quarters of all local calls were made for social reasons to family and friends. Another survey suggested almost 50 percent talked on the telephone to friends or relatives every day (Fischer 1992, 226).

ANXIETY, SECURITY, AND PRIVACY

Moving away from thinking of social impacts in broad sociological terms there have also been a number of studies which have considered the possibility of more personal, largely psychologically impacts of the telephone. From a more negative perspective there have been suggestions that the telephone may have added to the levels of anxiety about security and privacy and the general pace of domestic life. From a more positive perspective there have been suggestions that the telephone may have made people feel more secure and more connected.

A number of commentators have suggested that the telephone has contributed to more-tense-less-private household environments. What has been of particular interest is the possibility for the telephone to breach the traditional barriers between the public sphere and the private world of the household. Coupled with the massive growth of telephone directories, barriers between the individual and the outside world can be seen to be progressively diminishing. There is a long list of possible scenarios where telephone interactions can breach privacy and lead to anxieties: Callers can telephone at any time; be unseen; misrepresent their identity; demand immediate responses, catching the receiver unprepared; be verbally intimidating; and solicit for unwanted products and services. Receivers can become anxious as they await anticipated calls that do not arrive or, alternatively, callers can become anxious when their calls remain unanswered. Those who have had previous experience of calls at unexpected times bringing bad news can harbor fearful associations with the sound of all unanticipated calls. Parents can become anxious because they don't know to whom their children are talking. Many of these scenarios for telephone-related anxieties are intuitively plausible. These themes of the telephone as a potentially menacing invasion into the home have also appeared at different times in broader cultural representations of the telephone in fiction and film (Lubar 1993, 139–141; Stern and Gwathmey 1994, 94–95).

While these negative scenarios clearly exist, there is some evidence that, for most users, telephone related anxieties are not taken as serious enough to outweigh the benefits of the telephone. As noted earlier, the bulk of the literature reports that women, who are likely to spend more time in the domestic sphere than men, have generally responded to the telephone in a positive way helping to maintain social bonds and overcome feelings of isolation. It is also something of a truism that teenagers in particular have adopted the telephone to help them maintain contact with friends during a period of their lives when social bonding is a major preoccupation but when anxieties about physical appearances, privacy, and face-to-face communication are common (Stern and Gwathmey 1994, 103–116; de Sola Pool 1983, 132–133). The mobile telephone would appear to have filled this traditional niche and expanded it further.

Broader surveys have also suggested that while telephone users did find breaches of privacy annoying, that with the exception of calls in the middle of evening, the level of anxiety and irritation was not extremely great. Other surveys of users who lost their usual telephone service reported its absence made things feel less hectic but lead to feelings of uneasiness, isolation, and loss of control. These findings suggest that despite the annoying features of the telephone its potentials to help maintain contact with the outside world and provide opportunities to communicate more readily have encouraged many people to feel less anxious and more secure but in a domestic sphere that is potentially less private and busier (de Sola Pool 1983, 139–140; Fischer 1992, 246–247).

7

Stormy Weather, Telecommunications Deregulation, and the New Digital World, 1970s–

In the decades following the Second World War, firms and government interested in commercially exploiting new communications technologies began to raise questions about what role the *Bell* telephone monopoly should play in their development. In the 1950s firms attempting to enter into forms of microwave broadcasting challenged *Bell's* protected position. There was continued lobbying for firms to have the right to be able to operate private microwave systems. *Bell* countered these suggestions arguing that these systems would compromise its development and maintenance of the public network (Faulhaber 1987, 24–25).

Bell's status would also be challenged by various electronic firms who wanted to market terminal equipment. The most important of these was the 1968 "*Carterphone* case" where a Texas entrepreneur won the right for customers to attach Carterphone instruments to AT&T's lines. Carter filed an antitrust case against *Bell* when it threatened to refuse service to customers who used his instruments. *Bell* had appealed to the FCC claiming that allowing customers to use non-*Bell* equipment compromised the quality of the network as a whole and was not in the public interest. *Bell* resisted suggestions that as long as strict technical standards were applied to "foreign attachments" the integrity of the telecommunications system could still be maintained. They argued that there were no appropriate institutions in place capable of monitoring compliance (Faulhaber 1987, 27–30). The FCC enrolled the National Academy of Sciences (NAS) to consider the matter. The NAS disagreed with Bell's assessment and suggested that monitoring compliance could be done. This meant that by the mid-1970s, in theory, Bell's competitors could attach some of their equipment to the Bell system as long as it passed the FCC standards. The effects of these changes were not to be major as Bell adopted a defensive posture slowing down the adoption of new technology by encouraging lengthy debates about standards as a way of impeding the processes of monitoring compliance (Faulhaber 19878, 30). It would not be till Bell's final breakup in the early 1980s that more substantial changes would take place.

As a consequence of increased use of computers for data processing, and also of electronic office equipment such as fax and telex, and modems (devices that convert the digital signal produced by a computer into an analog form so it can be sent down a traditional telephone to another computer), large businesses and government were becoming increasingly dependent for coordinating things, like cash flows, investments, and production, on the rapid flow of huge amounts of digital information that traveled along telephone lines. The costs of communications and paying for telephone services were becoming an increasingly significant part of their budgets. The notion that telephone monopolies such as Bell were the most efficient ways to deliver these services would increasingly come under challenge (Reinecke and Schultz 1983, 79–98).

THE DIVESTITURE OF THE BELL SYSTEM

In the early 1970s Bell still controlled approximately 90 percent of the U.S. telephone service, but their grip was slipping. In 1974, in an act that symbolically marked the beginning of the end of the traditional organization of the telephone system, the U.S. Justice Department filed an anti-trust suit revisiting their longstanding concerns that it was inappropriate for AT&T and Western Electric to be part of the same company. This case would drag on for more than a decade. The first judge presiding over the case died and hundreds of millions of dollars were devoted to legal fees. Finally on January 8, 1982, Bell agreed to split up its operations.

In arrangements legally consented to on January 1, 1984, AT&T retained control of Western Electric and maintained an interest in long-distance operations. Its major concession was to divest itself of local operating companies. These local operations would be taken over by the seven so-called "Baby Bells." The "Baby Bells" would operate independently, handle local calls, and be able to move into the emerging cell phone market, but would be restricted in their involvement in phone equipment manufacture and long-distance services. These new arrangements also removed restrictions to other companies providing telephone services (Lubar 1993, 142).

The breakup of the *Bell* system represented the end for one of the longest standing technological and business systems in history. Around the time of divestiture in 1983 the *Bell* system had revenues of $65 billion, 1 million employees, assets worth $150 billion, and 84 million customers (Forrester 1987, 87).

While representing a radical departure from the past the divestiture of *Bell* would not constitute a simple clean new model for the operation of telephones; for instance, by 1985, the FCC gave AT&T permission to market office automation services, an area in the original agreement AT&T had been restricted from entering. Charles Brown, president of AT&T, highlighted the opportunities he believed divestiture would offer *Bell* and the consumer: "No one contemplated twenty-five years ago that a revolution in modern technology would largely erase the difference between computers and communications. As a consequence, the Bell system has been effectively prohibited from using the fruits of its own technology. And this new decree will wipe out these restrictions completely" (quoted in Lubar 1993, 142).

Alongside governments and businesses trying to work out ways to promote, invest in, or profit from new communication technologies, the divestiture of *Bell* also needs to be set against the stormy political climate of the time. One of the most significant set of political changes emerging from the late 1970s carrying into the 1980s were the widespread adoption of the economic policies of British Prime Minister Margaret Thatcher and U.S. President Ronald Reagan. Their policies would spark intense and ideologically charged debates about the appropriate role for economic regulation across much of the Western world. Both championed, in theory at least, the breakup of business and government monopolies, the need for as little government regulation of business as possible, and proposed arguments against "the spirit" of things like "universal service" and government-regulated or government-supplied services more generally. They believed it was better for the user to pay for services and let the economic marketplace sort out the most efficient way for services to be delivered.

INTERNATIONAL DEREGULATION OF TELECOMMUNICATIONS

While *Bell* had been a privately owned monopoly, and the norm in most other countries was for telephones to be managed as government publicly owned monopolies (PTTs), *Bell's* divestiture would still exert a wider international influence on the organization of telecommunications. Some nations, such as Japan, had directly modeled their government-owned service on the Bell system and, perhaps unsurprisingly, in the wake of *Bell's* divestiture, commenced (but very slowly) on a path to break up the monopoly of the Nippon Telegraph and Telephone Public Corporation (Forrester 1987, 94). For other nations who had not copied the *Bell* system directly, most still shared a number of *Bell's* features. Despite its private ownership, *Bell* had always been subject to significant government regulation, and "universal service" featured in most systems as an important goal. *Bell* had also provided a fairly obvious and more direct influence through being one of the leaders in developing telephone technology and international long-distance telephone services. What happened in the United States then would have both indirect and direct implications for the telephone systems of other countries especially those with financial linkages to the United States via transnational companies (Reinecke and Schultz 1983, 57–78).

The divestiture of *Bell* linked up with broader debates across the world about whether or not the public or private sector, or what mix, should control telephones and the political implications of breaking up such large traditional monopolies. In a 1983 article in *Business Week*, international telecommunications deregulation was described as ". . . a sticky problem for almost every government. Because the government-owned PTTs are such large employers, and are so heavily unionized, any attempt to turn them into competitive private companies invites a nasty political backlash from vast numbers of civil servants. What is more, the PTTs generally contribute healthy profits to government coffers" (quoted in Forrester 1985, 123).

The deregulation of the telephone system in the United Kingdom can be used to provide a brief case study of the broader international trends at the time of the divestiture of *Bell*. Between 1979 and 1984, as part of the controversial deregulation and privatization agenda of a newly elected Thatcher government, the U.K. telephone system, that had been a long-standing monopoly run by the Post Office, would be privatized. This was achieved by a number of steps. First, in 1979, the Post Office was split into two entities: Post and Telecommunications (British Telecom). In 1980, the Post Office monopoly on supplying telephone equipment, telephones, and PABXs was relinquished. In 1982, a new consortium called Mercury Communications was established that granted a license to build an alternative telephone network using fibreoptic cable in competition to British Telecom. In the same year the British government announced that it proposed to sell British Telecom. While not as big as its American cousin, its size and financial value displays the historical significance of its privatization. It was valued at 8 billion pounds sterling, and had 20 million customers and 240,000 employees. As could be expected the proposed sale generated heated public and industrial opposition. Concerns about job losses and increase in residential charges were raised. Government stuck to its plans and in November 1984, 51 percent of shares to British Telecom were sold and a regulatory body, Oftel (Office of Telecommunications), was established to oversee the changes to the British Telephone Industry (Forrester 1987, 93).

THE TELEPHONE IN THE INFORMATION SOCIETY

The divestiture of *Bell* and the commencement of the global restructuring of PTTs also has to be set against the much broader context of what many social theorists and various political commentators have described as the emergence in developed countries, such as Japan and Western Europe and the United States, of the "post-industrial" or "information society." Various social theorists, such as Daniel Bell (1974), suggested that through a variety of changes, but especially through the new potentials offered by computers and communications technology, the past focus of economic activity, culture, and employment around manufacturing industries would steadily move toward new knowledge-based industries involving the production, exchange, and consumption of information. This general theme of society undergoing a significant set of structural and technological changes emerging from the late twentieth century continues to the present, with numerous studies now focussing attention on the recent growth of the Internet and other "new media" (Flew 2005).

While there have been cautionary and pessimistic appraisals of the implications of the emergence of "the information society," many commentaries on the emergence of the information society have displayed a strong utopian streak (Kling 1996, 40–58). Some of these utopian claims match extremely well the excitement generated by the telegraph of almost 150 years before. In 1981, Japanese social scientist and government consultant Yoneji Masuda predicted, "[I]f industrial society is a society in which people have affluent material consumption, the information society will be a society in which the cognitive creativity of individuals flourishes throughout . . . a society in which everyone pursues the possibilities of his or her own future. . . . [I]t will be global, in which multicentered voluntary communities of citizens participating voluntarily in shared goals and ideas flourish simultaneously throughout the world" (Masuda quoted in Forrester 1985, 626). Like the telegraph in the past, such technological utopianism appears rather shallow, and the historical lessons that problems such as war and poverty cannot be addressed by better communication and more "information" alone are too easily forgotten (Winner 1986, 98–121).

Aside from these macrocosmic issues involved with the information society a number of persistent, more specific, questions about the social impacts of digitization and telephones on day-to-day life have reappeared. The most important of these have surrounded whether, or not, there are new risks to privacy, greater unemployment in traditional telephone-related industries and services, and the possibility of new patterns of work.

PRIVACY AND DIGITAL SURVEILLANCE

From a positive perspective new telephone services such as caller ID, low-priced answering machines, and "voice mail" empower the receiver of telephone calls by helping them create something of a boundary between the outside world and the private sphere of the household. With the traditional telephone, callers could invade the private space of the household by making unsolicited calls and interrupt the temporal relations of the private sphere by calling at unexpected times and if they chose maintain anonymity. The receiver of the call may now choose to return a call, or not to, and, in many contexts, may be able to determine the caller's number. Simultaneity and spontaneity can be replaced by something more similar to the temporal quality of the traditional letter. This is even reflected in the title of "voice mail." It is interesting to note voice mail may be more popular among receivers of calls than those making them: 90 percent of callers in a 1992 survey found voice mail irritating (Lubar 1993, 143).

Digitization also offers growing opportunities for new forms of caller surveillance. Computers can now be used to analyze vast quantities of telephone call data in ways that were previously much more labor intensive and expensive. "Transactional information," records of calls to whom, and when, can be easily recovered; and "profiling" programs, allowing identification of particular patterns of calls, are used to monitor caller behavior. Employers, via PABX systems, can bar certain numbers from being called and track employees phone use (Reinecke and Schultz 1983, 94). The increased awareness of the ease with which telephone calls in certain contexts can be monitored may even slowly have an effect on users in shaping the types of conversations they have and whom they speak to on telephones.

UNEMPLOYMENT

Questions about whether or not the emergence of new information technologies has contributed to higher levels of unemployment across the developed world were raised with some intensity during the 1970s and 1980s. The decline in the number of jobs in telephone repair was a typical example of concern. In many contexts the introduction of digital switching technologies required far fewer technicians to provide maintenance and demanded new types of skills: for example, much of the diagnosis of telephone faults could now be conducted via a computer analysis at a central exchange (Reinecke and Shultz 1983, 87).

While it is probably of little consolation to those who lost jobs during the boom in microelectronics in the 1970s and 1980s, it is true that new employment in services, some of them spin-offs of IT, and the emergence of a knowledge economy, have been generated since (Flew 2005, 150–157). It is beyond the scope of this text to evaluate the complex questions of whether, or not, computers and IT generate employment, or unemployment, in the long term, or whether the quality of remaining, or new forms of work, are better or worse. But, whatever position is taken on these larger debates, it is undeniable that an important social reality of the late twentieth century has been the experience of dislocation of many workers whose jobs were made redundant, or radically altered, by patterns of economic deregulation of industries linked to the traditional telephone and the growth of new form of information technologies (Forrester and Morrison 1994, 193–226).

THE ELECTRONIC COTTAGE

Changes to information technology have also been linked to more radical predictions that the home will replace the traditional workplace. Some of these speculations share similarities to predictions made prior to the home computer and the Internet about the decentralizing potentials of the telephone. History has shown these predictions to be too simplistic. The traditional telephone enabled professionals like doctors to coordinate home visits, attend emergencies, and assisted some businesses to decentralize their operations, but also offered contrary possibilities by at the same time, assisting in the coordination of evermore centralized office blocks (de Sola Pool 1983, 41–49).

During the 1980s, predictions about the decentralizing potentials of information technologies were rekindled by arguments that the amalgamation of the telephone and home computer and other information technologies would result in the birth of the so-called "electronic cottage." The widely cited "futurologist" Alvin Toffler wrote in 1981, that this would represent as big a break to the fabric of day-to-day working life in the late twentieth century as the shift from the first pre-industrial cottage workshops to the factories of the city during the Industrial Revolution. Toffler breathlessly suggested, "it takes an act of courage to suggest that our biggest factories and office towers may within our life times stand half-empty, reduced to use as ghostly warehouses or converted into living space. Yet this is precisely what the new mode of production makes possible: a return to the cottage industry on a new higher, electronic basis, and with it a new emphasis on the home as the centre of society" (quoted in Toffler 1981, 204). While the expansion of the Internet over the last decade has enhanced these possibilities, there have as yet been nothing like the radical changes anticipated by Toffler. A number of surveys have suggested that many workers find it psychologically difficult to work from home and complain of problems of concentration, motivation and social isolation (Forrester 1991, 213–227). Rather than provide a simple replacement for the space of the traditional workplace, new information technologies would appear to have been much more active in promoting the growth of "supplementary work, and the 'office on the run,'" where work is done both in a physically appointed workplace and at home (Flew 2005, 151–152).

CONSUMERS, REGULATORS, AND DIGITAL CONVERGENCE

The divestiture of *Bell*, and the ongoing international pattern of the breakup of government-owned telephone monopolies (PTTs), has helped form a much more complex environment for telephone consumers and regulators over the last 2 decades. From the 1980s a wider variety of telephone instruments and services has become available. Putting to one side cell phone technology, which will be discussed in the next two chapters, there are now numerous options for fashionable handsets, cordless telephones, "voice mail," caller ID, answering machines, and "pay-per" services. While massive demand for fashionable telephone instruments such as Mickey Mouse handsets etc. has never really materialized, the popularity of some "pay-per" services has sparked the growth of major industries; "telephone sex" for example was a $2.4 billion industry by 1990 (Lubar 1993, 143).

As telephone companies have sought to exploit the possibilities of the convergence of telecommunications technologies there have been regular suggestions that telephone networks now need to be thought of in new ways and, in particular, rather than being considered in isolation, they should be thought of in terms of the place they occupy in national information infrastructures. These discussions are often framed in terms of ISDN (Integrated Service Digital Network) development. Telephone companies have attempted to expand their niche in these infrastructures, by either working on ways their traditional lines can be used to best carry digital information such as the Internet, or encourage the rewiring of telephone lines to make them more digital friendly such as by the expansion of the use of fibreoptic cables, cell phone networks, and forms of microwave and radio transmission.

Accompanying the increased diversity of telephone services and plans for developing national and global ISDNs has been a larger variation in quality, costs, and charges for those using standard telephones. As a generalization, because long-distance services are less likely to be used to subsidize local calls and because of the greater call-carrying capacity of fibreoptics, microwave relays and satellites, long-distance call services have become cheaper. What is less clear is whether or not the wide provision of cheap local calls has suffered, and whether, or not, ideals of "universal service" have fallen by the wayside (Lubar 1993, 143).

While digital convergence is often used as a way of describing the current communications environment, like the early days of the rhetoric of "universal service" of 100 years ago, the actual meaning of "digital convergence" is not as simple as it sounds, and users may in fact be faced with an experience of communication divergence. A simple indicator of the diversity of the current communication environment are business cards that now routinely need to list an increasing number of user addresses: postal, email, telephone, fax, mobile, and Web page (Mueller 1997, 186–191). At the time of writing this, diversity is growing with the Internet and the mobile telephone, offering alternatives to the traditional telephone.

8

The Global Mobile, 1980s–

Not since the adoption of the pocket watch has a technology been as quick in becoming so widespread in its use as the mobile telephone (Agar 2003, 3–5). A recent market study from Portio Research released in January 2006 predicted that 50 percent of the world's population will be using a mobile phone by the end of 2009, and by 2011 there will be 3.96 billion users (Cellular News, January 20, 2006). Many early projections of the growth of mobile telephones failed to estimate the speed and magnitude of the growth of the mobile. Some academic commentators in the 1980s suggested market penetration of about 20 percent. There were also some prediction for more rapid growth. Duane L. Huff, vice president in charge of cellular development at Bell Laboratories predicted in 1983 that in 20 years mobile communications would be "commonplace" and "a necessity for many" (quoted in Huff 1985, 137). Around the same time a consultant's report commissioned for AT&T suggested that the total cell phone market would be around 900,000 (Brown 2002, 3). Even these enthusiastic assessments fall short of current assessments of 2 billion users worldwide (International Telecommunication Union 2006).

While much of the initial scientific and technical know-how for mobiles emerged in the United States, the use of mobile telephones would rapidly expand first in Northern Europe, Japan, and Southeast Asia and, over the last decade, to the rest of the world. Currently the fastest rate of growth of mobile phones is in Africa with a predicted 265 million new users by the year 2011. The top growth market was India, just ahead of China, with a prediction of 1.06 billion subscribers by 2011, and in third place Brazil, Indonesia, and Nigeria. Reversing earlier trends of a relatively sluggish uptake of mobile telephones, relative to other developed countries, the United States is now predicted to have the sixth fastest growth of mobiles with 66 million new callers by 2011 (Cellular News, January 20, 2006).

The widespread international use of mobile phones is not their only global feature; their manufacture also reflects the flows of raw materials, labor, and information of the global economy. A typical mobile will reflect the influences of Scandinavian-industrial design; its electronic circuitry will have been built using North American, Japanese, and Northern European technological know-how; the capacitors will be constructed from a rare mineral known as tantalum, which was probably mined in the Congo or Australia; nickel in the battery is likely to have been extracted from mines in Chile; the plastic casing and the liquid in the liquid crystal display (LCD) refined from petroleum products from oil sources in the Gulf, North Sea, or Russia; the case moulded into shape in Taiwan; and the bits and pieces assembled in a variety of, most probably, low wage countries (Agar 2003, 6–15). While it is a globally manufactured product, the lion's share of the profits from its sale will go back to Europe and the United States. An unusual negative side effect of the growth of mobiles, reflecting the interconnectedness of the global economy, has been the political "fall out" of the huge increase in the demand for rare raw materials needed to build mobile phone capacitors. Most important of these is a material known as tantalum. One of the world's main sources of tantalum is the Democratic Republic of Congo that has been racked since the late 1990s by a civil war: one of the factors contributing to the civil war has been battles between various political factions for mineral rights to extract tantalum. As prices for the material have risen, so has the intensity of the conflict (Agar 2003, 13–14).

PLANTING THE SEEDS FOR THE MOBILE, CAR PHONES, AND RADIO

The forerunner to the mobile phone was the radio telegraph developed by Guglielmo Marconi (1874–1937) in the late nineteenth and early twentieth centuries. The radio telegraph was initially adopted for the demands of maritime and naval communication. In times of fog and over distances where visual signals were useless, wireless telegraphy was obviously of tremendous benefit. The potential benefits encouraged significant resources to be put into its development. Wireless telegraphy was complex and expensive and therefore for some time limited to large commercial and military users. The steady miniaturization of electrical components during the twentieth century and the slow but steady improvements in understanding antennas and theories of electromagnetism would see the development of radar, radio, and television, and also make the convergence of the telephone and radio technologies possible (Agar 2003, 11–12).

The early use of radiophones was limited by the size of batteries and other components: large heavy components meant telephones needed to be carried in a car, or on a ship. Another set of important challenges surrounded the problem of how to best use the available radio spectrum. Marconi's early technologies produced radio waves that spanned a wide part of the radio spectrum. Even as it became possible to better tune transmission signals, the management of the radio spectrum still presented constraints to the spread of radiophones. If a large number of phones existed, all using their own designated frequency, the radio spectrum would quickly become saturated with signals. These constraints encouraged early mobile communications to be restricted to military and police radio with designated parts of the spectrum for their use and no or limited link back to the broader landline telephone system.

Police forces in the United States, starting in Detroit, had begun to experiment with the use of radio in automobiles as early as the 1920s. In the following 2 decades this technology would be improved and spread into military applications. One company emerging in this period was Galvin Manufacturing Corporation, which soon changed its name to suit its product: Motorola. Motorola helped develop the "walkie talkie" and other portable radios that became important in World War II (Agar 2003, 35–36).

After the war, there were some efforts to develop commercial applications for these technologies. Private "ham radio" would become an important area, as were efforts to develop a mobile telephone highway service marketed for truckers and reporters. As a result of the research of Alton Dickieson and D. Mitchell from Bell Laboratories, in 1946, the first mobile telephone calls were made as part of a highway service. By 1948 the service spread along almost 100 cities and highway corridors and had 5,000 customers who made 30,000 weekly calls. The system faced various limitations: at most 3 subscribers could call at the same time in the same city, it cost $15 a month and 30–40 cents a local call, and the equipment needed to be carried in an automobile, as it weighed 80 pounds (AT&T 2006).

CELLULAR COMMUNICATION

Despite *Bell's* efforts to develop a mobile telephone highway service it would not be until the 1970s that U.S. regulators (and regulators in other developed countries) would start to address the management of the radiofrequency spectrum in a way that would encourage mobile telephones to become more than a niche technology. Around the same time that *Bell* launched its "highway service," engineers at Bell Laboratories had also begun to develop some of the most important technologies of the second half of the twentieth century, such as the transistor. Coinciding with the development of the transistor, other *Bell* scientists, D. H. Ring and W. R. Young, had begun to develop the principles of "cellular communication": a way of dividing the radiofrequency spectrum to avoid interference and allow for a greater number of signals/users. Ring's cellular proposal appeared in a Bell Laboratories Technical Memorandum, published on December 11, 1947; its title was "Mobile Telephony—Wide Area Coverage." Ring suggested that it should be possible to allocate a small number of frequencies to a pattern of hexagons (cells) in a given region. As users moved from one cell to another they could be allocated a different frequency as long as no one else was using the same frequency as a given user in one of the small cells at any one time, and as long as the first and last hexagons in a pattern were far enough apart not to offer interference, the pattern hexagons (cells) could be repeated across a larger region. This allowed for a large number of users of a relatively small part of the overall radiofrequency spectrum.

This repeated pattern of cells offered technological challenges in the form of developing ways to automatically switch through frequencies and link regions of cells together (Agar 2003, 19–22). Cellular communication would also require a huge investment in infrastructure, primarily a massive number of "base stations" that would receive and retransmit the relatively weak signals produced by mobile telephones. The need for a growing number of base stations, particularly as newer generations of mobile telephones are introduced, continues to the present day. This system would require ways of identifying and tracking individual phones and linking this new telephone system with the old. Switching technologies in the 1940s was not up to the task that a workable cellular system demanded, and

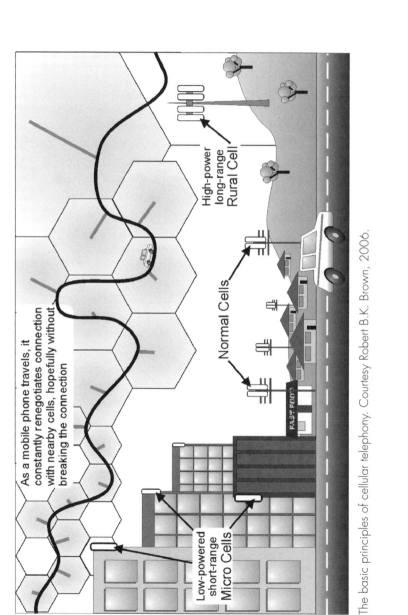

As a mobile phone travels, it constantly renegotiates connection with nearby cells, hopefully without breaking the connection

High-power long-range Rural Cell

Normal Cells

Low-powered short-range Micro Cells

The basic principles of cellular telephony. Courtesy Robert B.K. Brown, 2006.

understanding the best ways to the use the radiofrequency spectrum at higher frequencies was still being developed.

MOBILES, A GRUBBY RESEARCH BACKWATER?

There have also been some suggestions that engineers at Bell Laboratories were also distracted by the wealth of other technological possibilities of the time such as the ill-fated picturephone system. A National Science Foundation (NSF 1998) survey interviewed a number of scientists who had worked from the 1960s on areas relevant to the development of cell phones. These scientists reported that very few scientists were working in the area. One went as far as describing being involved in mobile radio as comparable with being "lost in the desert," and another commented that the field was treated like a "grubby backwater." The first major effort to put together a scientific and technical publication relevant to the development of cell phones did not occur until January 1979, when the *Bell System Technical Journal* devoted an entire issue to cell phones. Even then, the NSF suggested that *Bell* may have published to deter possible competitors moving into these areas of research rather than to seriously develop new technology (NSF 1998).

While these technical limitations help account for why Ring's idea was initially ignored for so many decades, it is also important to remember that the society of the late 1940s and 1950s was in many respects quite unlike the society that has grown since the late 1970s and 1980s where industries and consumers have grown to expect rapid changes in new communications technology (Agar 2003, 26). Various changes in models of legislation and regulation of traditional telephone services across many developed countries from the late 1970s to the 1980s (as described in Chapter 7) also played a role in providing a space (albeit, often a contradictory and volatile one) for the mobile to grow by destabilizing the way the telephone industry viewed "universal service," the needs of users, and the economics of the telecommunications.

DIFFERENT NATIONS AND PATTERNS OF DEVELOPMENT OF THE MOBILE

The story of this early period of the emergence of the mobile telephone also has a strong international dimension and provides an excellent case study on the way technical standards can influence patterns of technological innovation. The discussion below will provide a snapshot outlining the subtle and not so subtle national differences that have shaped the development of the mobile. It is important to remember that mobile technology is still changing and that any account of such recent history has not had the benefit of time for events and interpretations to be fully digested.

THE UNITED STATES

After lobbying from *Bell* in 1974 the FCC earmarked part of the radiofrequency spectrum for an experiment in cellular communication; in 1977 Illinois Bell (Chicago's operator for Bell) was authorized to install the first cellular telephone system of 10 base stations. The system began being used in late 1978. The system had a capacity of 2,000 users who could link up via telephones carried in their cars and base stations to the traditional telephone system. There has been some speculation that this legacy of the cell phone in the United States being originally conceived as automobile phone would be one of the factors that would later contribute to U.S. manufacturers falling behind relative to their European counterparts in efforts at miniaturization (Agar 2003, 43).

The FCC treated the experiment as a success and began to plan ways of spreading the system across the United States. Their plans were to be heavily influenced by the regulatory environment of the 1970s–1980s. By the late 1970s the *Bell* system was under challenge and the idea of government-protected monopolies politically unpopular. When the FCC began to grant licenses for cellular networks in 1984 it was by auction on a "city-by-city" basis in an environment where competition among telecommunication companies was being encouraged. There was so much interest in and so many applications for licenses that the FCC found that it was more difficult to manage the granting of licenses than they had anticipated. This lead them to decide that after the first thirty licenses were granted for the 30 biggest U.S. cities they would conduct a lottery for the other two-thirds of major cities. This "initiative" further widened the number and variety of firms seeking licenses. Even though the United States did adopt a standard analog system for the phone to communicate with a base station ("AMPS" Advanced Mobile Phone System) but the system in general was poorly coordinated. Even as companies such as *Bell* began to reconsolidate in the 1990s the United States would still have an unwieldy system characterized by numerous small companies working on a "city-by-city" basis (Agar 2003, 39–41). The United States would also be a number of years behind Europe in moving from the analog (AMPS) system to digital systems.

Digital systems have advantages in being able to transmit more than just voice and help increase the capacity of calls that systems can handle. It was not until the late 1980s that the United States would adopt technically sophisticated digital systems, but again their implementation would be disorganized (Agar 2003, 68–69). The cell phone would really only take off in the United States relative to most of Western Europe during the late 1990s (Agar 2003, 31–43).

A number of explanations have been put forward to account for the United States being initially slow in developing the mobile:

i. As noted above, there was a patchwork of different standards that encouraged a similar diversity of technical developments. While many of these mainly analog systems were technically efficient, in their own terms and region, their potential growth was limited because of lack of coordination. If the mobile was to satisfy its full potential as a universal and mobile technology, better standards to allow users to communicate from one system, and from one place to another, needed to be developed.

ii. The United States may have been a victim of its own early success in developing paging devices that became popular in business contexts and that offered an effective alternative to mobile use.

iii. The early, small but successful U.S. market niche, of mobile phones operating from automobiles may have slowed down efforts to miniaturize components.

iv. There was an initial reluctance for U.S. companies to implement calling party pays (CPP) billing. CPP billing systems charge the person initiating a call. These systems quickly became common outside the United States. In the United States the mobile user paid a fee to accept an incoming call. This helps explain that in 1998 only 20 percent of U.S. mobile users had given their number to more than 10 people (Robbins and Turner 2002, 80–93).

In all, while it had been U.S. technological ingenuity that had helped give birth to the mobile phone, the United States would not initially build on this early success and take the lead in developing and refining the mobile; instead this would take place in Scandinavia, and later in Europe through the initiatives of the European Union.

SCANDINAVIA

There were a number of social and economic conditions in Scandinavian countries that would help shape and nurture the development of the mobile telephone. Sweden is a country with dense forests and spread-out population. So a basic system of mobile radiophones had been operating from the 1950s. The significant starting point to the growth of the Scandinavian mobile phone industry can be identified as 1967 through the initiatives of the chief engineer of Swedish Telecom Radio, Carl-Gosta Asdal (Agar 2003, 49). Asdal suggested that Sweden should develop an automated mobile telephone network integrated with the landline network. Further studies supervised at Sweden's Telecom radio laboratories began testing Asdal's ideas. By 1969 this project expanded to enlist the support of engineers from other Nordic countries: Denmark, Norway, and Finland. this became the Nordic Mobile Telephone Group.

At this time these countries were characterized by a tradition that recognized the importance of negotiation and consensus. Sweden, for instance, had been a leading country in experimenting with models of consultation between labor, industry, and government in the introduction of new technologies (the industrial relations strategies of Volvo and the car industry are well-known examples). The emergence of Scandinavian countries as important players in the early mobile phone industry highlights a point sometimes made by economists specializing in the study of technological change: while competition can help drive technological change, a lack of predictable standards and too much diversity can also make it difficult for designs to be refined, business systems to operate efficiently, and for users and predictable markets to develop. Exploiting their traditions of negotiation and consensus in developing new technologies, Scandinavian countries initiated things like customer surveys and encouraged communication between governments and experts and communication between engineers themselves.

Out of this socio-technical environment the NMT (Nordic Mobile Telephone) standard was developed. The radiofrequency spectrum was thought of as a national resource that needed to be carefully managed. In 1981 the NMT system was launched and by 1986 the NMT system was so popular that its capacity was full and a second higher frequency system NMT 900 was introduced (Agar 2003, 49–50).

The pooling of expertise and common standard helped give the Scandinavian countries an important early lead with the development of mobile telephone technology. By 1987 approximately 2 percent of the population of Nordic countries were mobile telephone subscribers. Other European countries were impressed with the Nordic model and Spain, the Netherlands, Austria, and Belgium attempted to adopt NMT systems although the success of this adoption was restricted by price and did not turn out to be as rapid as was the case in Nordic countries. The successful emergence of powerful mobile telephone companies from Scandinavian countries provide a good example of how technological change is not driven by technologies or markets alone, and highlights the importance of standard setting and regulatory cultures (Agar 2003, 44–51).

THE EUROPEAN UNION AND THE EMERGENCE OF GSM

Viewing the emerging Nordic success, larger European countries such as France, Germany, Italy, and Britain decided it was time to develop their own systems. Compared to the Scandinavian systems these early European systems were nowhere near as successful and Europe, through the 1980s, suffered from similar problems to the United States in offering a patchwork of different standards and different mobile telephone systems. Many engineers and administrators were aware of the possible benefits of developing a more harmonious Europe-wide system. These beliefs became linked to some important broader political ideologies and debates being played out within the European Union.

While nation-states such as France and Britain have often been ambivalent about how much they should be integrated into the EU, there has been continuing counter arguments emphasizing the importance of the EU as a unified market and also an engine for developing new technologies. It has often been suggested that only a unified Europe would have the wealth and know-how to compete with technical and industrial giants such as the United States and Japan. It was also argued that for the EU to be economically viable it needed to be culturally viable and for this to occur not only would physical barriers/borders between member states need to be reduced but also barriers between communication. Within this framework of pro-EU sentiment, meetings were held in Stockholm in 1982 between engineers and administrators from 11 European countries; they had gathered to consider the development of a Europe-wide so-called GSM system. GSM stood for 'Groupe Speciale Mobile,' which later changed to Global System for Mobile Communications. This was a digital system, and came to be known as the second generation of mobile telephones, as it would come to replace the first analog generation. Being a digital system also meant that it would be able to offer the possibility of providing not just voice but other information services.

By 1987 GSM prototypes had been experimented with and most political differences sorted out. EU bureaucracies promoting the development of GSM made their philosophy for developing the standard explicit. In recommendations issued by the Council of the European Communities of June 25, 1987, 87/371/EEC, it is noted: "A coordinated policy for the introduction of a pan European cellular digital mobile radio service will make possible the establishment of a European market in mobile and portable terminals which will be capable of creating by virtue of its size, the necessary development conditions to enable undertakings established in Community countries to maintain and improve their presence on world markets" (quoted in Agar 2003, 60).

By 1991 the GSM system began operation, and by 1995 most of Europe was covered. By 1996 GSM systems were operating in 103 countries (Agar 2003, 62–63). GSM was not necessarily the best system in technical terms but once it became established it allowed manufacturers to focus on incremental improvements in things like transmission technologies and handsets and also provide the space for the developments of less immediately commercially relevant but technically interesting innovations such as SMS or text messaging (Trosby 2004, 187).

There were also quirky legal issues that initially limited the number of firms who would enter into the initial development of GSM. The EU insisted that manufactures of elements of the GSM system would need to indemnify themselves against future possible risks of patent litigation. The GSM system relied on a number of small intricate innovations ("microinventions") and some manufacturers, particularly from the United States and Japan, were unwilling to take the legal risks of patent infringements if they engaged in further refining GSM. Mobile phone giants such as Nokia of Finland, Ericsson of Sweden, and the United States's Motorola, were nevertheless not deterred, and embarked on intense and profitable competition to develop smaller and smaller handsets and steadily improve the components making up the system (Agar 2003, 56–66).

The GSM system, incorporating ongoing improvements has become the most popular "platform" for mobile telephones globally. The global trade GSM association founded in 1987 whose goals are to promote the interests of GSM operators throughout the world boasts on its Web site: "That at the end of September 2005, it consisted of more than 675 second and third generation mobile operators and more than 150 manufacturers and suppliers. The Associations members provide mobile services approaching 1.65 billion customers across more than 210 countries and territories around the world"(GSMtmWORLD, 2006).

THE UNITED KINGDOM

The patterns of growth of the mobile phone system in the United Kingdom during the 1980s was strongly influenced by debates about privatization and deregulation that raged from the late 1970s. From more or less the beginning of the history of the telephone, the industry had been run as a monopoly by the British Post Office. This would change in 1981 when the Post Office passed on telephone operations to a newly created public corporation, British Telecom (see Chapter 7). The British Government decided to promote the development of their own mobile phone system. They advertised that two licenses to run analog systems would be offered. One of these would be taken up by a company partnership between British Telecom and the security firm Securicor, which operated under the name of Cellnet; the second license was granted to a consortium of defence electronic company Racal and Millicom that had operated cell phone systems in the United States. They would operate under the name of Vodaphone. Initial profits for Cellnet and Vodaphone were rather disappointing, but as the mid-1990s approached, further licenses were granted leading to another two operators entering the market: "One 2 One" and Orange. The entry of these two new operators would lead to intense competition, advertising, new marketing strategies, and the introduction of digital networks. By the late 1990s, the United Kingdom had one of the world's highest rates of mobile phone use and its mobile phone companies had become industrial giants (Agar 2003, 70–89).

JAPAN

Japan is correctly recognized as one of the leaders in consumer electronics and a country where consumers have shown a willingness and enthusiasm to embrace new technologies. Like other developed countries, the first commercial mobile phone services would begin to appear in the 1970s (some of the world's earliest services were started by Nippon Telephone and Telegraph in 1979 around Tokyo and Osaka). While Japanese technology companies were providing some of the components for cellular systems in other countries, by the late 1980s mobile phone use was still minimal in Japan. Japan had also found it difficult to break into Europe's GSM-dominated system and had been deterred by the regulatory complexities of the United States and the United Kingdom.

This situation was to change in the late 1980s to early 1990s when the favorable ingredients of competition, but within an organized framework of standards, emerged. Three consortia: Nissan, NTT, and Japan Telecom would be granted licenses to operate within a new Japanese mobile standard. The 1993 launch of the Japanese digital system, marks an amazing success story in the adoption of mobile phones. Market leader NTT would boast an increase from one million users in 1993 to 40 million by 2002. One of the interesting features of NTT's system was its promotion of so-called "I" mode. This would come to be known as DoCoMo. "I" mode allows phone users, for a fee charged to their phone bill, to access a selection of various forms of digital information (Agar 2003, 94–101). The DoCoMo system, by linking mobiles to other digital information sources, has anticipated the current development of the so-called third generation (3G) of Mobile Telephones.

3G OR NOT 3G? THE FUTURE OF THE MOBILE

The 1990s can be seen as the era of the consolidation of the mobile telephone as a commonplace technology. But as the current millennium unfolds the industry looks to be becoming yet again volatile. While the spread of mobiles in developing countries, and the beginning of more rapid growth in the United States, still represents a huge profitable industry, many major mobile companies have expressed anxieties about sustaining their profits.

With awareness that markets will become saturated without further innovations a number of European mobile phone companies paid up to 80 billion pounds sterling in the late 1990s in licensing fees to operate a so-called new "third generation" (3G) of mobile telephones (Burgess 2004, 52). The 3G is based on the idea that mobile telephones should be able to integrate with, and even substitute for, the functions performed by personal computers. Watching videos and TV, accessing the Internet, engaging in "e" commerce and "e"mail, would in theory all become possible via the mobile. Phone users would "enjoy" "constant contact" an "always on" connection. The 3G services use higher frequencies than other mobile systems. Because such frequencies do not travel as strongly as lower ones, some commentators have calculated up to three times as many antennas will be required for the service to properly operate (Burgess 2004, 52).

At the time of writing, there have been some anxieties that 3G technology is not growing as quickly as planned, partially because incremental innovations to 2G mobile technology would appear to be able to offer many of the same services. These are sometimes described as 2.5 G mobile technologies. Despite the huge expansion of mobiles across the world, documented at the beginning of this chapter, and the promise of 3G, many mobile telephone companies entered the new millennium with economic anxieties. The huge penetration of mobile telephones in to day-to-day life had provided huge profits, but lucrative markets in developing countries risked becoming saturated. For example, in 2001, Nokia, Ericsson, and Motorola all indicated concern with their profits falling behind anticipated levels; and, in 2002, Vodaphone registered a major loss (Burgess 2004, 53).

Users would appear to constitute part of the problem for mobile companies. Callers exploiting "top up" plans (avoiding long-term phone contracts) and text messaging, while still lucrative, are not generating the profits many firms desire. In some ways users have not turned out to be as predictable as industry had hoped. For instance, some surveys have suggested that the mere purchase of a mobile telephone does not mean that it will be frequently used. Similar to the 1930s when numerous U.S. farmers temporarily stopped using their telephones, in more recent times, many who purchase a mobile, use it rarely, or buy the cheapest pre-paid mobile telephone package and leave their mobile turned off, only to be activated during emergencies. One of Europe's largest telephone operators, Orange, announced on April 9, 2002, that 750,000 of its prepaying customers had neither received, nor made a call, in the last 3 months (Burgess 2004, 44). Mobile companies will be banking on users, particularly the booming youth market embracing 3G technology, and are becoming increasingly self-conscious to try to anticipate whether or not users will embrace the new services that are being offered.

9

Mobile Cultures, 1990s–

Assessing the social impacts of any technology is a demanding task and, as was noted in earlier chapters on the standard telephone, something that needs to be done with caution to avoid merely "reading off" the social impacts from the logical possibilities offered by the technology in question. It also must be kept in mind that as a new technology it may not yet have a stable form. It is also likely that some of the social "impacts" are "novelty effects" where there are tensions between old ways of doing things and new and more visible technology. An interesting example of this can be found in studies that have revealed the widespread "use" of fake or dummy mobile phones in places such as Budapest and Chile in the late 1990s. A Chilean newspaper reported that in a police "crackdown" on drivers talking on their mobiles, a third of those stopped were actually using fake mobiles (Persson 2001, 2). It is hard to imagine merely being seen to be using a mobile telephone would carry the same incentive in Nordic countries where almost the entire population now uses them. As mobiles become more popular it could be expected that they will become less visible as "technological mediums." Perhaps, in the not too distant future, they will become like the traditional telephone, where most users are so familiar they become almost invisible as a technology (Cooper 2002, 20–21).

MOBILES MEAN BUSINESS

An interesting similarity between the early days of the mobile and the early days of the standard telephone is that many analysts anticipated that businesses would be the main users of the telephone. Well-known historian of the telephone, Ithiel de Sola Pool noted in an 1983 article "Will Mobile Telephones Move" ". . . the desire for personal conversation is not the main test of the prospective significance of mobile telephones. Their role in increasing business productivity is likely to be far more important than their role in casual conversations" (quoted in de Sola Pool 1985, 145).

These assessments are perhaps unsurprising when the high costs of early mobiles are considered. In 1984 Motorola released its first commercial mobile telephone with a suggested price range of $3,000 to $4,000 (NSF 1998, 10).

The bulky phones which appeared in the 1980s were often both business and status symbols. In the popular 1987 film *Wall Street*, which satirized the corporate greed of the 1980s, numerous scenes feature the tireless businessman Gordon Gekko, played by the actor Michael Douglas, shouting orders, at all times and places, down his brick-like cell phone (Agar 2003, 144). Using his phone may well have doubled as a part of his fitness regime: state-of-the-art phones in 1987 were 700–800 grams or about 2 pounds in weight, and these represented a marked improvement on the first commercially launched Nokia Mobira Talkman that weighed 4.8 kilograms or about 10 pounds 9 ounces (Burgess 2004, 39).

As sales grew, incentives to develop smaller phones followed. Promoters also picked up on business themes using advertisements that had marked similarities to those of the standard telephone from 70 years before. In 1986 instructions were issued to British Telecom mobile cell phone salespeople under the heading "Turning Idle time to Productive Time." The following advertising banter followed: "When you're away from your office and your phone, you're effectively out of touch with your business. You can't be contacted. Nor can you easily make contact yourself. Take a mobile telephone—a cell phone—with you and you get a double benefit. You're totally in touch, ready to take instant advantage of business opportunities when and where they occur. And you can make maximum effective use of "dead-time"—time spent traveling—turning it in to genuine productive hours" (Agar 2003, 83). Like much earlier telephone advertising, the ability to act at a distance, give orders, and coordinate business were promoted as benefits of using a mobile but tellingly, in addition, these advertisements also link the mobile to time and travel.

The two themes, of organizing time and mobility, have become increasingly important for the promotion of the mobile. In a little over 10 years the mobile moved from being considered mainly as a business tool, to a symbol of technological progress and a cultural and practical necessity, particularly for young adults. Taking a quick snapshot of the mobile in popular culture just over a decade later, *Wall Street's* businessman Gekko, who uses his weighty mobile to give orders, has been replaced in the 1999 film *The Matrix* (geared toward a young adult audience) by the lead player Neo played by Keanu Reeves, who uses his sleek "state-of-the-art" Nokia 8110i mobile to communicate with his team of heroes on their travels between their more enlightened world and an evil mechanical one.

One of Nokia's marketing managers Heikka Norta proudly announced on the day *Matrix* was launched, "Nokia's mobile phones create the vital link between the dream world and the reality in *Matrix*. The heroes of the movie could not do their job and save the world without the seamless connectivity provided by Nokia's mobile phones. Even though our everyday tasks and duties may be less than those of the heroes of *Matrix*, today we can all appreciate the new dimension of life enabled by mobile telephony. As the leading brand in mobile communications, Nokia is proud to see that the makers of *Matrix* have chosen Nokia's mobile phones to be used in their film" (quoted in Agar 2003, 146–147). By 1999 mobile telephones were being packaged not just as a business tool but a necessity to function in the modern world and even as a medium to allow users to enter into new ones.

MOBILES AND YOUTH CULTURE

While business uses and convenience and cost still feature significantly in the promotion of mobile telephones the targeting of younger people as one of the key markets for mobile telephones has continued to expand. As noted earlier, the standard telephone was a technology already popular with teenagers feeding off their needs for peer contact and sociability. It is unsurprising that some of these uses would spill over into demand for mobile telephones, particularly when these age groups also tend to have, in wealthier countries at least, disposable income for fashion and socializing.

The popularity of mobile phones among younger people cannot be underestimated. In a number of surveys young people report that the mobile telephone is one of their favorite possessions, with younger teenagers valuing their mobile phones more than older ones (Campbell 2005, 3).

Youthful enthusiasm for mobiles is reflected in high levels of youth mobile phone ownership across most developed countries. Comparing a sample of the results of a variety of surveys from the late 1990s these trends are clearly shown. Ownership levels reached; 80 percent of 13–20-year-olds in Norway in 1999; 90 percent of people under 16 in the United Kingdom in 2001; 56 percent of children aged 9–10 in Italy in 2003; and, 33 percent of children aged 10–13- and 43 percent of 13–15-year-olds in Australia in 2003 (Campbell 2005, 3). In the United States, cell phone use is on the rise after a slower start than Northern Europe, Japan, and Southeast Asia, but similar general patterns apply. In the United States, in February 2002, 13 percent of 12- to 14-year-olds had cell phones, and by December 2004, the figure had risen to 40 percent, with 14 percent 10- to 11-year-olds owning cell phones. Surveys in 2005 estimate 16 million teens and younger had cell phones in the United States (Petrecca 2005). A 2005 survey has suggested the situation has now been reached where in many developed countries adolescents are more likely to own a cell phone than their parents do.

Some detractors of the boom in the use of mobiles by young people have suggested many become so obsessed that the time they spend on the telephone is contributing to an epidemic of youth obesity. On the positive side there have been some studies that have speculated that youthful obsessions with mobiles may be replacing some of social and emotional functions previously satisfied by smoking (Burgess 2004, 62).

Manufacturers have become increasingly interested in working out ways to maximize their profits by encouraging young people to use cell phones. In 2005, in the United States, Mattell toy company licensed a range of "My Scene" cell phones aimed at the market of children 12 years and under; Walt Disney followed with similar marketing strategies. Some critics, such as Gary Ruskin executive director of the Commercial Alert Advocacy Group, have cautioned that these marketing strategies risk exposing children to targeted advertising campaigns, potentially pressuring them to buy accessories and ring tones. One of the most interesting recent developments for children has been the release of the so-called Firefly mobile. It is a phone designed specifically for 8- to 12-year-olds. Some of its design features anticipate and restrict some of the uses teenagers often favor in mobile phones. For example, it comes with only five keys; parents use a private PIN number to construct the limited 22 outgoing numbers the phone is programmed to call; it has "mom" and "dad" speed dials; does not allow text messaging; has no camera; no Internet access; and can be programmed to receive calls from a limited set of numbers. On November 15, 2005, the Firefly won the CES innovation award, presented by the Consumer Electronics Association, and the 2006 International Consumer Electronics Association (Petrecca 2005).

On the other side of the Atlantic, keen to maintain and boost the young cell phone user market, companies such as BT Cellnet, Mercury, One2one, Orange, Vodaphone, and Ericsson have begun to sponsor ethnographic fieldwork studies on teenage mobile telephone users with the aim of gathering ideas about how they might incorporate new features into mobile telephones to increase their capture of these markets (Berg, Taylor, and Harper 2003, 433).

From a positive perspective this work can be seen to follow the tradition of industrial design pioneered by engineers such as Henry Dreyfus in the 1930s, to understand users and design for them the technologies that best satisfy their needs. From a more sceptical perspective it can be seen as a way of studying users, so that manufacturers can attempt to create new "needs" that can basically be satisfied with variations of existing technological "know-how". Coupled with slick advertising campaigns, desire to purchase new variations of products and use new services can be continuously stimulated. Nevertheless, as the longer history of the telephone has suggested, even in a fast-paced modern world of media manipulation the potential unruliness of users cannot be overlooked and the paths that technologies take may not always be so simple to predict for either promoters or detractors.

UNRULY USERS: FASHION, BOMBS, AND TEXT MESSAGES

There are a number of examples where users of mobile telephones have been "unruly" using them in ways that designers did not originally intend. The most benign example would be the way mobile telephones, for many younger users in particular, have taken on various symbolic and fashion roles; originally mobiles took on standard appearances as functional devices, with limited variations in appearance or choice of "ring tone." A user could display their status by owning a new or expensive model, or placing the telephone in a pouch, or craft some kind of decorative covering (Plant 2002). Studies by British anthropologists have also suggested that it is not just young women who use their mobiles to express fashion and identity. In a study published in 2000 some young men in bars appeared to use their phones as "lekking" devices (devices used by birds such as preening their feathers before mating) to try to attract female attention (Lycett and Dunbar 2000, 93–104).

Some manufacturers moved quickly to consider the merging of fashion with function. Nokia, for example, aggressively marketed mobiles from

1999 with slim designs without a protruding antenna, and on to which different colored Facia (casings) could be added (the Nokia 3210). Given Nokia's success with variations on this basic design, it can be seen a little like the Dreyfus "300" traditional telephone design of the 1930s that would set a design benchmark for further telephones (see Chapter 5 and Agar 2003, 113–121). In recent years, a whole youth-orientated industry has emerged where subscribers can rent "ring tones" featuring everything from extracts from their favorite pop songs to the demented ramblings of fantasy cartoon frogs: various games are a standard addition to most mobiles. From a more sinister perspective, unruly mobile telephone users have also been able to reengineer them to help act as detonators for terrorist bombs. Authorities in Bali and in London have on recent occasions, as a measure to stop bombs being detonated, temporarily disabled their mobile telephone networks.

The most significant example of users influencing designers and promoters to think differently about mobiles has nevertheless been with the immense global popularity of Short Message Service (SMS) or text messaging. The general idea of text messaging was discussed by GSM planners in Europe in the mid-1980s. It was thought that it might be a useful way of alerting telephone users of incoming messages. Apocryphal stories of the history of the mobile also suggest texting was only added as an afterthought because of leftover space on the telephone's computer chip. Various paging systems already existed which were similar. So the assumption was users might find texting useful but not revolutionary. In particular, manufacturers didn't anticipate the popularity and wide range of applications of texting (Trosby 2004, 193).

Since the first SMS text messages were sent in the early 1990s (Agar 2003, 105–110) texting has become a massive global phenomenon, becoming hugely popular in Southeast Asian countries (such as Singapore and the Phillipines) with Europe, China, and Australia following closely behind. Similar patterns have also occurred in Japan, where a slightly different technological system NTT DoCoMo is used to fulfill similar functions. Calculations have been made that in 2004, 500 billion text messages were being sent per annum, at a rate of close to 100 text messages per person in the world. In 2001, 250 billion short messages were sent, whereas only 17 million in 2000; in China alone 18 billion text messages were sent in 2001. Interestingly, the United States has been slower in adopting text messaging with only 13 messages sent per average user in 2003 (Wikipedia 2006, for more detailed statistics see, Cellular Online 2006). This may have been the result of the United States offering cheaper rates per minute for standard calls than other nations, greater disposable wealth of younger users with less constraints to their "mobile budget," and a lack of uniform relevant technical standards in the United States making texting difficult to do across different telephone networks. Texting is currently on the rise in the United States through cultural exposure to it via "text" voting on immensely popular "reality" TV programs such as *American Idol*.

The popularity of text messaging was initially with less-wealthy users, helping explain its huge initial growth in Southeast Asian countries. Texting was also associated with the growth of the option to buy prepaid telephone cards, where the user has a set number of calls, can monitor how much they are spending, and have the ability to "top up" their capacity as needs arise. These systems of payment emerged as an alternative to the variety of (normally more expensive) package deals favored by manufacturers popular in the early days of the mobile, where users paid by monthly accounts (or bills). The growth of these measured forms of prepaid billing have been noted by the OECD as something that has helped to globally stimulate faster growth of cell phones (Burgess 2004, 34). It has also helped contribute to the growth of younger users who may be legally ineligible or discouraged by parents from buying into contract plans. Top-up systems generally charge more per call but the user has more flexibility and does not need to guarantee ongoing payments. For the prepaid mobile user, text messages offer a cheap measurable way to maintain social contacts.

Moving beyond what were originally price considerations text messaging has now taken on a variety of cultural forms. Abbreviated styles of writing, constructed phonetically so as to take up less space, and use less phone time have started to resemble their own dialects. Attesting to the popularity of texting the most popular Christmas book in the United Kingdom, in 2000, was a book on the language of texting (Burgess 2004, 44).

Ethnographic studies of texting among teenagers have also begun to trace some of the features of teenage text culture. They have observed trends such as certain text messages being treated as having sentimental value (suggesting that future phones should make them easier to be stored); that text messages were often shared between small groups of users with a phone being passed around a small group of friends to show off a displayed message; communication involved an etiquette of prompt answers—anything longer than 15–30 minutes for a reply incurred the need for an apology; and, that texting allowed social contact to be maintained with a minimal amount of time and effort (Berg, Taylor, and Harper 2003, 435). A more worrying spin-off from texting has been the concern that it may be particularly well suited to antisocial exchanges. Bullying by text can take place at any time and the target may read the words repeatedly, and, unlike a verbal taunt, some young people suffering text bullying may find it a form of abuse that they perceive is more difficult to escape from as it is written and may take on a more lasting quality in their memory (Campbell 2005, 5).

Texting has also encouraged the "rebirth" of an old technology, in a new guise: Morse code. There have been recent speed contests between experienced Morse code operators and text users. Morse code is apparently much faster to text with. Firms, such as Nokia, have shown some interest in developing mobiles that can convert incoming Morse code to text; or building phones that may emit light pulsed in Morse-code-readable with a camera phone. In Nokia's patent application they describe it as a new communication channel that does not pollute the Rf band (Dybwad 2005).

Texting would also appear to be well disposed to organizing spontaneous "public gatherings." Cheap, brief messages can be treated like chain letters and a common message be quickly circulated to broad numbers of users, sometimes with anonymity afforded to the source. Spontaneous gatherings brought together via texting have the potential to enhance grassroots protests and democracy, as well as criminally inclined riots.

The most famous example of the positive power of texting, was the use of mobiles and text messaging in the Philippines in the early twenty-first century to help overthrow a poorly performing undemocratic government. In poorer countries, such as the Philippines, the costs of land lines had denied much of the population easy access to telecommunications. Mobiles were able to tap into a new market of less-wealthy users who could use cheap, limited prepaid "top-up" cards. While individual calls were expensive, sending brief text messages were cheaper and allowed individuals to maintain contact with each other despite their tight budgets; "top-up" cards also allowed users a greater degree of anonymity. By 1996, 10 percent of the population owned a mobile telephone. In 1998, Joseph Estrada was elected President: it did not take long for serious concerns to be raised that he was falling into the bad habits of government corruption that had plagued the Philippines for decades under the Marcos regime.

During the year 2000, public support for Estrada plummeted. Traditionally, government in the Philippines would have been able to suppress dissent in the traditional mass media, but this time they found themselves unable to "silence" a chorus of text messages demanding Estrada be removed. At the peak of the protest over 100 million text messages were being sent each day. Some helped facilitate rallies that put huge pressure on the Estrada government. Estrada was finally toppled and a new president, Gloria Macapagal-Arroyo, came to power in 2001. Aware of the role that texting had played in the downfall of her predecessor, she attempted, in the face of considerable protest, to ban malicious, obscene, and profane texts, and proposed taxing text messages (Agar 2003, 105–110).

The political implications of texting may also constitute a "double-edged" sword. Whereas the tale of Estrada suggests that texting can be a vehicle to encourage grassroots democracy, it can also be used as a tool to help provoke and coordinate riots with undemocratic aims. The circulation of text messages was recently blamed as one of the factors helping to promote race riots in Cronulla, a suburb of Sydney, Australia, in December 2005 (AsiaMedia 2005). Racist political activists exploited the possibilities of mobiles and email to help amplify growing ethnic tensions and "turf rivalries" between teenage gangs and help "organize" riots. They sent out hate "email" and chain text messages calling for a party to be held to celebrate "national identity" at a beachside suburb. Police were completely taken by surprise when the crowd quickly swelled to more than 5,000 (some had traveled for miles to attend). Fueled by alcohol, the "party" quickly went out of control and mobs of youths rampaged down streets chanting racist slogans and attacking people of "Middle Eastern" appearance, as well as the police. In retaliation, later that evening, opposing gangs of youths drove through various Sydney suburbs vandalizing cars and properties, and performing random attacks of violence on people of "Anglo" appearance. While the retaliation involved far fewer youths than the original riots, it too, was coordinated via text messaging and mobile phones. One of the tools police used to track down "ring leaders" in the months that followed was to trace "hate" text messages and mobile and Internet use.

While the mobile phone did not cause the downfall of Estrada, or the race riots in Sydney, in both cases it did offer a means to amplify political sentiments and allowed swift mobilization of people in ways that were difficult for authorities to anticipate or control. These things would have been much more difficult to do with traditional communication technologies.

MOBILES AND HEALTH

Concerns with disease being spread by bacteria developing in telephone mouthpieces, warnings about the use of telephones during thunderstorms, and phobia's about electricity, have intermittently arisen during the life of the traditional telephone. While some studies have suggested that mouthpieces do, in fact, house bacteria and that using the telephone during electrical storms is indeed dangerous, the traditional telephone never attracted wide concern in relation to its negative impact on health (de Sola Pool 1983, 99). The short history of the mobile has been quite different.

The mobile emits radiofrequency radiation from its antenna as it "communicates" with a base station which, in turn, also emits radiation in

signaling to other phones, towers, and back. The profusion of devices emit-
ting radiofrequency radiation constitutes one of the problems engineers deal
with in allocating frequencies to different devices and making sure there is
no interference. The evergrowing number of these devices also means that
many of us now lead our lives effectively bathed in a weak electromagnetic
"soup." Most scientists suggest that the levels the community are exposed
to are generally low and not worthy of major concern; alternatively, some
others see it as an unknown new form of electrical "smog" with possible
long-term public health implications.

Health concerns linked to mobiles have been related to both radiation
exposures to the user, and community exposures to those living near cell
towers. For users of mobiles, there have been worries that even if a mobile
produces relatively low levels of radiation, much of it is being absorbed in
the user's head which is often, more or less, in contact with the phone. For
those living in close proximity to towers (who are experiencing levels of
radiation less than a mobile) concerns have mainly focused on the fact that
these people are being continuously exposed.

Concerns about cell towers are often dismissed as the by-product of
the so-called NIMBY (Not In My Backyard) syndrome. People simply do
not want cell phone antennae in their backyards whether they constitute a
health hazard or not, and anxieties of individuals using mobiles are rejected
as an example of scientific ignorance and paranoia. Some critics have gone
as far as suggesting these responses are prime examples of people rejecting
modernity and being unable to rationally deal with technological change.
These styles of argument normally focus on the fact that there has been
considerable scientific work done by government and official scientific
agencies, such as the International Committee for Non-Ionising Radiation
Protection (ICNIRP) of the World Health Organization (WHO), which
have largely dismissed the health implications of exposure to low levels of
EMF (Chapman and Wutzke 1998, 614; Burgess 2004).

Dismissing health concerns out of hand nevertheless overlooks the fact
that most of these government reports have still acknowledged that there
are scientific uncertainties, and gaps in knowledge, and not all reports have
dealt with these uncertainties in the same way (Mercer 1998, 291–294).
A good example of the latter can be found in comparing the recent report
The Health Council of the Netherlands (2002) that, more or less, rejects
all possible risks out of hand, with the U.K. Stewart Report (2000), which
discourages excessive use of mobile phones by children, pending further
research. Most importantly, health surveys (epidemiological studies) of users
of mobile telephones may take many years to come up with results as
brain tumors can take decades to arise (Graham-Rowe 2003, 12–13). The

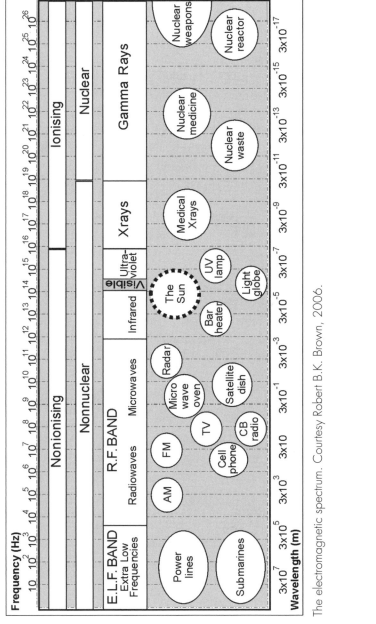

The electromagnetic spectrum. Courtesy Robert B.K. Brown, 2006.

huge, and very recent, use of mobiles by children and teenagers has also raised questions about whether these groups may have greater sensitivity to biological damage from exposure to low levels of radiofrequency radiation than adults.

The safety assurances of organizations such as the WHO, while capturing the views of a majority of scientists, have also been undermined in the eyes of some critics, who suggest that they have been unnecessarily conservative in their scientific assessments, routinely failing to consider the work of a significant minority of scientists who have expressed concerns about possible risks to health of radiation associated with cell phone use (Slesin 2005). There have also been a small but persistent number of scientists who have taken these criticisms further and claimed that the massive financial influence of the electrical and cell phone industry has been used to suppress information and studies that suggest cell phones may be a health risk (Carlo and Schram 2001, Maisch 2006). When pressed by these types of accusations, government authorities, the mobile industry, and organizations such as the WHO have regularly responded by suggesting that most of the science suggesting possible risks is uncertain or of poor quality (Repacholi 2005).

The issue becomes yet again more complex, when it is considered that the cell phone health debate doesn't stand alone but hooks into a longer-standing scientific controversy involving theoretical disagreements about the ways radiofrequency radiation and electric and magnetic fields interact with living things (Steneck 1984; Mercer 2001, 84). Since the 1970s some scientists have been concerned that even very low levels of long-term radiation exposure from microwaves, radiofrequencies (telecommunication devices, for example, radio and cell phones, and electric and magnetic fields (powerlines and electrical infrastructure) may constitute a long-term health risk through causing changes to brain chemistry and interrupting other biological processes. Other scientists have disagreed, suggesting that these forms of radiation (nonionizing radiation) can only cause immediate health problems and only then, when exposures are sufficiently strong to electrocute or burn people (induced current and thermal effects). The radiation exposure experienced by mobile phone users and people living near cell towers is generally well below levels which can cause immediate biological effects. But some scientists caution that possibility of less immediately visible long-term health effects can't be dismissed.

In keeping with the controversial background to the question, the matter has found its way into the courts (Grasso 1998). The most important recent case has been *Newman v Motorola* 2003. Christopher Newman, a

Baltimore Neurologist, claimed his use of a Motorola analog phone between 1992 and 1998 caused the development of a brain tumor behind his right ear. The case was watched closely by the cell phone industry as it was the first of a dozen or more cases pending against them involving mobile telephone personal injury claims. Estimates of possible industry liabilities were in excess of $6 billion. The case was ultimately heard in the United States District Court for Maryland where Newman's claims were dismissed (Edmond and Mercer 2004, 239–243). The favorable outcome for the mobile industry has discouraged further litigation for the immediate future but given ongoing scientific uncertainties. It is unlikely that this will be the last time the question will be considered by courts.

Despite reasonably widespread media reporting of the cell phone "health debate," the sale of mobiles, as noted earlier in this chapter, would not appear to have been adversely affected. This suggests that mobile users perceive that the benefits of the mobile outweigh any possible health risks.

"HULLO ... WHERE ARE YOU?" SOCIAL SPACE AND THE MOBILE

One of the most important features that sets the mobile telephone apart from the traditional telephone is quite literally its "mobility." The traditional telephone was linked to specific place, a home or business address. The mobile is not linked to a place but rather a person. When a traditional telephone is answered the standard response is "hullo," callers and receivers may identify themselves, and the caller knows where the person answering is located in space. In calls using mobiles, after saying "hullo," opening exchanges frequently involve talk about place and space: "I'm calling from x" "on the way to y" or "where are you calling from." Locating a call has practical consequences for mobile users; it can help provide callers with an idea about what might be an appropriate or inappropriate style or topic of conversation depending on bystander audiences. By being attached to a person, rather than a place, the mobile can intensify some of the tendencies displayed with the traditional telephone in relation to breaching the spaces between the private and public spheres. Whereas the traditional telephone affected the space of the home by providing contact, both wanted and unwanted, with the outside world, the mobile has the capacity to blur these categories further by bringing the private sphere into the public sphere. Private and personal telephone conversations can take place in almost any public setting, and, depending on the sense of etiquette of the mobile user, involve almost any topic.

Some analysts have suggested that the willingness for some mobile callers to engage in what may have traditionally been conceived of as private conversation in such public settings may reflect important shifts in the way public space, and even broader community, is conceived. People using mobile phones while traveling in trains, walking, or driving automobiles through city landscapes can, in a sense, insulate themselves from the physical space they are in and take their own social world with them, communicating with immediate friends and family but disengage from those physically around them. This combination of engagement/disengagement allows users to become detached observers of the public spaces through which they travel. These possibilities of social disengagement fit in with the concerns of some social theorists that mobile telephones and other new technologies, such as the Internet, may encourage "cyberbalkanization" where people can limit their social interaction exclusively to those who share similar interests" (Burgess 2004, 64).

Other commentators have sought to explain why many people are able to have private conversations in public places on mobiles without encountering more objections (although codes for the use of mobiles in theatres and public places are steadily evolving). Drawing on the famous sociologist Irving Goffman, who studied the way people function in everyday life, the mobile can be seen as an "involvement shield." Traditional "involvement shields" in public settings were things like newspapers, or books, which one cannot literally hide behind, but through engagement with, present themselves as inaccessible. Being inaccessible while conversing about private matters in earshot of others, requires some intricate social negotiations; the listeners act as if they are not hearing (even though they can) by adopting a posture of "civil inattentiveness." To suggest how quickly these social codes evolve it is interesting to observe listeners finding it initially more difficult to maintain their "civil inattentiveness" when a caller is using a hands free set which may be a less familiar "involvement shield" than even the tiniest of mobiles (Persson 2001, 2).

This detachment of engagement with the immediate space a person is communicating from on the mobile does nevertheless have its physical (as well as social limits). The growing evidence of mobile telephones contributing to automobile accidents provides one indication. Another, perhaps more poignant, example can be drawn from considering the role of the mobile in the tragic unfolding events of September 11, 2001. People on upper floors of the World Trade Center, aware that they were unlikely to survive, made heart wrenching mobile phone calls to loved ones to bid them farewell; passengers on United Airlines Flight 93, hearing, via accounts gleaned from conversations on their mobiles, of the hijackers intentions to use their aircraft as a weapon, made a courageous but unsuccessful attempt to take over the aircraft.

PRIVACY SECURITY AND ANXIETY

Apart from effecting the way public space is viewed and blurring of public versus private in terms of the rules of conversation, the possibility of users being more or less continuously in contact has a number of implications for the way privacy and security are viewed (Marx 1994). From a privacy perspective the user is increasingly vulnerable to being monitored in relation to whom they make calls, how often and where they make them from. This "transactional information" is easy to gather digitally as part of it is necessary for the purposes of charging or billing calls. Accessing the records of mobile users has increasingly become a feature of police investigations into crime.

The fact that mobile systems can be used to identify the specific location of the user has captured the vision of advertisers who can automatically send promotional messages to users as they approach various shops or amenities. There has been some concern that the same technologies could allow police or governments to track users without their consent. The same possibilities have also been promoted as a way of increasing security, particularly of children. Children can wear a mobile that allows them to be monitored by their parents, everywhere from playing in the park to visiting friends. Despite parents believing that such technologically mediated surveillance of their children enhances their children's safety, a Finnish study noted a downside that while children's use of mobiles allowed parents to keep a better track of their children's physical whereabouts, they were no longer as well informed about their children's friends or peer group (Burgess 2004, 64).

This possibility for continuous contact, more intense than the era of traditional telephone, has inspired some critics to note a tendency for some users to become more or less addicted to their mobiles and there have been reports of high levels of anxiety experienced by some users if they cannot have access to a mobile. There have also been concerns that the psychology of constant contact may be substituting quantity of trivial superficial communication at the expense of quality and considered communication that may require more patient "face-to-face" contact. Alternatively, other

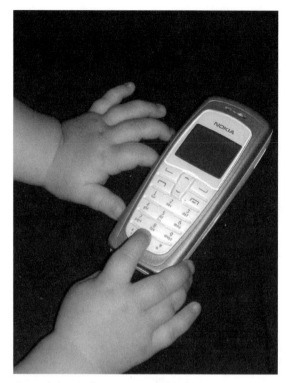

The mobile telephone. A powerful electric toy?
Photograph by Andrew Phillips, 2006. Used with
permission.

commentators have suggested that mobiles augment existing ways of communicating, and that effects on space and time and styles of communication will vary according to different social contexts and users (Green 2002, 290–291). As mobiles become increasingly familiar, and more sociological studies are done, the answers to some of these questions should become clearer.

Glossary

AMPS (Advanced Mobile Phone System). An early U.S. analog mobile telephone system.

Audion. Invented by Lee de Forest in 1906. The audion consisted of three parts—a vacuum tube that had inside it a filament which emitted electrons when heated, a positively charged metal plate which attracted electrons, and a negatively charged grid. The grid controlled the flow of electrons between the filament and the plate. Applying a signal to the grid modulated the current and produced an amplified signal in the plate circuit. The audion was modified to create *the high vacuum thermionic tube* or *repeater*. These devices were used to amplify signals as they passed along long-distance telephone lines.

Cellular communications. In 1947 D. H. Ring from the Bell Laboratories suggested that it was possible to allocate a small number of frequencies to a pattern of hexagons (cells) in a given region. As a user moved from one cell to another they could be allocated a different frequency, as long as no-one else was using the same frequency as a given user in one of the small cells at any one time, and as long as the first and last hexagons in a pattern were far enough apart not to offer interference, the pattern hexagons (cells) could be repeated across a larger region. This system allows a large number of callers to use a relatively small part of the overall radiofrequency spectrum. These principles, linked to powerful computer systems for switching and coding and decoding of signals, have provided the basis for the development of the mobile telephone network (see line drawing in Chapter 8 in the text).

Contact pressure transmitters. In the late 1870s Berliner and Edison worked on models for the telephone transmitter that worked much more effectively than Bell's original designs and were the forerunners to the idea of the microphone. They were based on the idea that sound waves could lead to the pressure between electrodes that were in continuous contact to be varied. These changes in pressure would in turn increase or diminish the resistance of a circuit, producing electrical undulations in the circuit similar in form to sound waves. Edison improved these transmitters by placing carbon granules between the electrodes (see line drawing of "standard" telephone in the text in Chapter 6).

DoCoMo. Japanese digital mobile telephone system developed during the mid-1990s.

Fibreoptics. An alternative to traditional copper cables developed in the early 1970s. In this system, information was carried by modulated light through glass cables rather than by electrons.

GSM (Global System for Mobile Communications). The world's most popular technical platform for mobile telephones developed in the European Union in the 1980s. GSM originally stood for "Groupe Speciale Mobile."

Information theory. One of the key figures in the development of information theory was Claude Shannon of the Bell Laboratories. In 1948 he wrote the "Mathematical Theory of Communication." One of Shannon's aims was to determine the most efficient ways a message could be sent along a channel with as little distortion as possible. From this work it was realized that there were a number of ways that the "information content" of an original message at the sender's end could be radically compressed and minimized and still be reconstructed meaningfully at the receiving end of a channel. Developing increasingly better ways of coding and decoding information, such as turning information into a **digital** form (ones and zeros) as is done with digital computers, means the quality of the transmission of a signal becomes far less important than in traditional analog communication systems.

ISDN (Integrated Service Digital Network). The notion that telephone networks shouldn't be considered in isolation but rather in terms of the place they occupy in national information infrastructures.

Liquid variable resistance transmitter. Elisha Gray and Alexander Graham Bell experimented with transmitting devices that involved a diaphragm which responded to vibrations made by the voice. The diaphragm had a wire connected to it that was dipped in a slightly acidic solution connected to an electric circuit. In response to the vibrations of the voice the wire would dip either deeper or shallower in the solution and in turn increase or decrease the resistance of the electrical circuit.

Loading coils. Devices invented in the 1890s by George Campbell and Michael Pupin. They were small electromagnets that, by being placed at regular intervals along a telephone line, helped maintain the strength of a telephone signal as it traveled along a cable.

NMT (Nordic Mobile Telephone standard). One of the first mobile telephone standards developed in 1981 for Sweden, Denmark, Norway, and Finland.

PABX. Private Automatic Branch Exchange

Page effect. Effect named after the work in 1837 of American physicist William Charles Page. Page explored the possibilities of producing sounds by rapidly magnetizing and demagnetizing metal rods. The sounds that they emitted displayed a relationship to the rate at which the rod was magnetized/demagnetized.

Post-industrial or information society. The idea becoming popular among many social theorists from the late 1970 that, through a variety of changes, but especially through the potentials offered by computers and communications technology, the past focus of economic activity, culture, and employment around manufacturing industries would steadily move toward new knowledge-based industries involving the production, exchange, and consumption of information.

SMS (Short Message Service, text messaging or texting). The general idea of text messaging was discussed by GSM planners in Europe in the mid-1980s. It was thought that it might be a useful way of alerting telephone users of incoming messages. At the time manufacturers didn't anticipate that texting would become a huge phenomenon.

Transistor. Invented in 1948 by William Shockley, John Bardeen, and Walter Brattain of the Bell Laboratories. The transistor was the forerunner to the microchip which has allowed the continuous miniaturization and increase in the power of computers. Transistors (from the idea of transit-resister) function as miniature switches by controlling the amount of current that can flow between two terminals by a voltage applied to a third terminal. Using different transistors, electrical circuits can be put together far more compactly and with greater reliability and durability than traditional switches and vacuum tubes.

Bibliography

Agar, John. *Global Touch: A Global History of the Mobile Phone.* Cambridge, MA: Icon Books Ltd., 2003.

Aronsen, Sidney H. "Bell's Electrical Toy: What's the Use? The Sociology of Early Telephone Usage," in Ithiel de Sola Pool (ed.), *The Social Impact of the Telephone*, pp. 15–39. Cambridge, MA: The MIT Press, 1977.

AsiaMedia. "Australia: SMS Calls for Race Riots out in Four States," UCLA Asia Institute. http://www.asiamedia.ucla.edu/article-pacificislands (December 15, 2005).

AT&T. "First Mobile Telephone Call," *AT&T Labs-Innovation-Technology Timeline.* http://www.att/attlabs/reputation/timeline/46mobile.html (accessed March 31, 2006).

Bargellini, Pier L. "An Engineer's Review of Antonio Meucci's Work in the Invention of the Telephone," *Technology in Society*, 15: 409–421, 1993.

Bektas, Yakup. "Cultural Constructions of Ottoman Telegraphy, 1847–1880," *Technology and Culture*, 41: 669–696, October 2000.

Bell, Daniel. *Coming of Post Industrial Society*, New York: Basic Books, 1974.

Berg, Sara, Alex Taylor, and Richard Harper. "Mobile Phones for the Next Generation: Device Designs for Teenagers," *CHI 2003*, 5(1): 433–440, April 5–10, 2003.

Briggs, Asa. "The Pleasure Telephone: A Chapter in the Prehistory of the Media," in Ithiel de Sola Pool (ed.), *The Social Impact of the Telephone*, pp. 40–65. Cambridge, MA: The MIT Press, 1977.

Brown, Barry. "Studying the Use of Mobile Technology," in Barry Brown, Nicola Green, and Richard Harper (eds.), *Wireless World: Social and Interactional Aspects of the Mobile Age*, pp. 3–15. London: Springer-Verlag London Ltd., 2002.

Bruce, Robert. *Bell: Alexander Bell and the Conquest of Solitude*. Boston, MA: Little Brown & Company, 1973.

Burgess, Adam. *Cellular Phones, Public Fears, and a Culture of Precaution*. Cambridge, UK: Cambridge University Press, 2004.

Business Week (October 24, 1983). "Telecommunications Liberalization," in Tom Forrester (ed.), *The Information Technology Revolution*, pp. 120–136. Oxford, UK: Basil Blackwell Ltd., 1985.

Campbell, Marlyn. "The Impact of the Mobile Phone on Young People's Social Life." Paper presented to a conference on Social Change in the 21st Century on 28th October 2005 at Queensland University of Technology. C. Bailey and K. Barnett (eds.), *Social Change in the 21st Century 2005 Conference Proceedings*. http://www.socialchange.qut.edu.au/conferences/socialchange/2005proceedings.jsp (accessed March 31, 2006).

Carlo, George and Martin Schram. *Cell Phones: Invisible Hazards in a Wireless Age*. New York: Carroll & Graf, 2001.

Carlson, Bernhard. "The Telephone as a Political Instrument: Gardiner Hubbard and the Formation of the Middle Class in America, 1875–1880," in Michael Trad Allen and Gabrielle Hecht (eds.), *Technologies of Power: Essays in Honor of Thomas Parke Hughes and Agatha Chipley Hughes*, pp. 25–56. Cambridge, MA: The MIT Press, 2001.

Cellular News. "Half the World Will Use a Mobile Phone by 2009." http://www.cellular-news.com/story/15674.php (January 20, 2006).

Cellular Online. http://www.cellular.co.za/index.htm (accessed March 31, 2006).

Chant, Colin (ed.). *Science Technology and Everyday Life 1870–1950*. London: Routledge & Open University, 1989.

Chapman, Simon and Sonia Wutzke. "Community Panics about Mobile Phone Towers," *Australian and New Zealand Journal of Public Health*, 21(6): 614–620, 1997.

Coe, Lewis. *The Telephone and Its Several Inventors*. Jefferson, NC: McFarland & Company, Inc. Publishers, 1995.

Cooper, Geoff. "The Mutable Mobile: Social Theory in the Wireless World" in Barry Brown, Nicola Green, and Richard Harper (eds.), *Wireless World: Social and Interactional Aspects of the Mobile Age*, pp. 20–21. London: Springer-Verlag London Ltd., 2002.

Dreyfus, Henry. *Designing for People*. New York: Allsworth Press, 2003.

Dybwad, Barb. Patent Highlights. http//www. lasers, optics and photonics resources and news-optics.org (March 12, 2005).

Edmond, Gary and David Mercer. "Daubert and the Exclusionary Ethos: The Convergence of Corporate and Judicial Attitudes to the Admissibility of Expert Evidence in Tort Litigation," *Law and Policy*, 26(2): 231–258, April 2004.

Farley, Tom. *Private line.* http://www.privateline.com. A Tom Farley Production. West Sacramento, CA, 2006. (accessed March 31, 2006).

Faulhaber, Gerald R. *Telecommunications in Turmoil: Technology and Public Policy.* Cambridge, MA: Ballinger Publishing Company, 1987.

Fischer, Claude. "Touch Someone: The Telephone Industry Discovers Sociability," *Technology and Culture,* 29(1): 32–61, January 1988.

Fischer, Claude. *America Calling: A Social History of the Telephone.* Berkeley, CA: University of California Press, 1992.

Flew, Terry. *New Media: An Introduction* (2nd edn.). Melbourne, Australia: Oxford University Press, 2005.

Flichy, Patrice. *Dynamics of Modern Communication: The Shaping of Modern Communication.* London: Sage, 1995.

Forrester, Tom. *The Information Technology Revolution.* Oxford, UK: Basil Blackwell Ltd., 1985.

Forrester, Tom. *High Tech Society.* Oxford, UK: Basil Blackwell Ltd., 1987.

Forrester, Tom. "The Myth of the Electronic Cottage," in Tom Forrester (ed.), *Computers in the Human Context* (2nd edn.), pp. 213–227. Cambridge, MA: The MIT Press, 1991.

Forrester, Tom and Perry Morrison. *Computer Ethics: Cautionary Tales and Ethical Dilemmas in Computing* (2nd edn.). Cambridge, MA: The MIT Press, 1994.

Galambos, Louis. "Theodore N. Vail and the Role of Innovation in the Modern Bell System," *Business History Review,* 66: 95–126, Spring 1992.

Gorman, Michael. "Alexander Graham Bell's Path to the Telephone," Technology, Culture & Communications. SEAS, University of Virginia. 1994. http://www3.iath.virginia.edu/albell/homepage.html (accessed June 5, 2006).

Graham-Rowe, Duncan. "Special Report: Mobile Phone Safety," *New Scientist,* (179): 12–13, 2003.

Grasso, Laura. "Cellular Telephones and the Potential Hazards of Rf Radiation: Responses to the Fear and Controversy," *Virginia Journal of Law and Technology,* 3 (2), 1998. http://www.vjolt.net/archives.php?issue=3 (accessed June 5, 2006).

Green, Nicola. "On the Move; Technology, Mobility, and the Mediation of Social Time and Space," *The Information Society,* 18: 281–292, 2002.

Green, Venus. "Goodbye Central: Automation and the Decline of 'Personal Service' in the Bell System, 1878-1921," *Technology and Culture,* 36(4): 912–949, October 1995.

Grosvenor, Edwin S. and Morgan Wesson. *Alexander Graham Bell: The Life and Times of the Man Who Invented the Telephone.* New York: Harry N. Abrams, 1997.

GSMTMWORLD. "About GSM Association." http://www.gsmworld.com/index.shtml (accessed March 31, 2006).

Health Council of the Netherlands. Mobile Telephones: An Evaluation of Health Effects. Publication No. 2002/01E. The Hague: Health Council of the Netherlands, 2002.

Heap, Nick, Ray Thomas, Geoff Einon, Robin Mason, and Hughie Mackay (eds.). *Information Technology and Society*. London: Sage, Open University, 1995.

Hellman, Hal. *Great Feuds in Technology: Ten of theLiveliest Disputes Ever*. Hobokin, NJ: John Wiley and Sons, Inc., 2004.

Hempstead, Colin A. "Representations of Transatlantic Telegraphy," *Engineering Science and Education Journal*, 18–25, December 1995.

Hoddeson, Lillian. "The Emergence of Basic Research in the Bell Telephone System, 1875–1915," *Technology and Culture*, 22(3): 512–544, July 1981.

Hounshell, David. "Elisha Gray and the Telephone: On the Disadvantages of Being an Expert," *Technology and Culture*, 16(2): 133–161, April 1975.

Huff, Duane L. "The Magic of Cellular Radio," in Tom Forrester (ed.), *The Information Technology Revolution*, pp. 137–146. Oxford, UK: Basil Blackwell Ltd., 1985.

International Telecommunication Union. "ICT Statistics," http://www.itu.int/ITU-D/ict/statistics/ (accessed March 31, 2006).

James, W. Carey. *Communication as Culture: Essays on Media and Society*. New York: Routledge, 1989.

John, Richard R. "The Politics of Innovation." *Daedalus*, 127(4): 187–214, Fall 1998.

Katz, James E. and Mark A. Aakhus (eds.). *Perpetual Contact: Mobile Communication, Private Talk, Public Performance*. Cambridge, UK: Cambridge University Press, 2002.

Kennedy, Robert C. "Cartoon of the Day: *A Candid Opinion of the Submarine Telegraph*," (Commentary on cartoon originally published, May 16, 1857 by Frank Bellew) *Harpweek* LLC, New York, 2005. http://www.harpweek.com/09Cartoon/BrowseByDateCartoon.asp?Month=May&Date=16 (accessed March 31, 2006).

Kline, Ronald. "Resisting Consumer Technology in Rural America: The Telephone and Electrification," in Nellie Oudeshorn and Trevor Pinch (eds.), *How Users Matter: The Co-construction of Users and Technologies*, pp. 51–66. Cambridge, MA: The MIT Press, 2003.

Kling, Rob. "Hopes and Horrors: Technological Utopianism and Anti-Utopianism Narratives of Computerization," in Rob Kling (ed.), *Computerization and Controversy: Value Conflicts and Social Choices* (2nd edn.), pp. 40–58. San Diego, CA: Academic Press, 1996.

Lipartito, Kenneth. "When Women Were Switches: Technology, Work, and Gender in the Telephone Industry, 1890–1920," *American Historical Review*, 99(4): 1075–1111, October 1994.

Lubar, Steven. *Infoculture: The Smithsonian Book of Information Age Inventions*. Boston, MA: Houghton Mifflin, 1993.

Lycett, J. and R. Dunbar. "Mobile Phones as Lekking Devices among Human Males," *Human Nature*, 11(1): 93–104, 2000.

Maddox, Brenda. "Women and the Switchboard," in Ithiel de Sola Pool (ed.), *The Social Impact of the Telephone*, pp. 262–280. Cambridge MA: The MIT Press, 1977.

Maisch, Don. "*EMFacts Consultancy,*" Lindisfarne, Tasmania, Australia. http://www.emfacts.com (accessed March 31, 2006).

Martin, Michele. "Communication and Social Forms; The Development of the Telephone 1876–1920," *Antipode*, 23(3): 307–333, July 1991.

Marvin, Carolyn. *When Old Technologies Were New: Thinking about Electric Communication in the Late Nineteenth Century.* Oxford, UK: Oxford University Press, 1988.

Marx, Gary. "New Telecommunications Technologies and Emergent Norms," in Gerald M Platt and Chad Gordon (eds.), *Self, Collective Behaviour and Society: Essays in Honour of Ralph Turner.* Greenwich, CT: JAI Press, 1994. http://web.mit.edu/gtmarx/www/telecom.html (accessed June 5, 2006).

Masuda, Yoneji. "Computopia," in Tom Forrester (ed.). *The Information Technology Revolution,* pp. 620–647. Oxford, UK: Basil Blackwell Ltd., 1985.

McLuhan, Marshall. *Understanding Media: The Extensions of Man.* New York: Mentor, 1964.

Mercer, David. "The Hazards of Decontextualised Accounts of Public Perceptions of Radiofrequency Radiation (RFR) Risk," *Australian and New Zealand Journal of Public Health*, 22: 291–294, 1998.

Mercer, David. "Overcoming Regulatory Fear of Public Perceptions of Mobile Phone Health Risks," *Radiation Protection in Australasia*, 18(2): 84–94, 2001.

Meyer, Ralph O. *Old Time Telephones: Technology Restoration and Repair.* New York: TAB Books, Division of McGraw Hill, Inc., 1995.

Moore, James. "Communications," Chapt. 7, pp. 200–250, and "Everyday Life and the Dynamics of Technological Change," Chapt. 1, pp. 9–40, in Colin Chant (ed.), *Science, Technology and Everyday Life 1870–1950.* London: Routledge & Open University, 1989.

Moyal, Anne. "The Feminine Culture of the Telephone: People, Patterns and Policy," in Nick Heap, Ray Thomas, Geoff Einon, Robin Mason, and Hughie Mackie (eds.), *Information Technology and Society*, pp. 284–310. London: Sage, Open University, 1995.

Mueller, Milton L. *Universal Service: Competition, Interconnection, and Monopoly in the Making of the American Telephone System.* Cambridge, MA: The MIT Press, 1997.

Noakes, Richard J. "Telegraphy Is an Occult Art: Cromwell Fleetwood Varley and the Diffusion of Electricity to the Other World," *British Journal for the History of Science*, 32: 421–59, December 1999.

NSF. The role of NSF's Support of Engineering in Enabling Technological Innovation-Phase II, "Chapter 4: The Cellular Telephone: SRI Policy Division Science Technology and Economic Development," 1998. http://www.sri.com/policy/csted/reports/sandt/techin2/chp4.html (accessed June 5, 2006).

Nye, David E. "Shaping Communication Networks; Telegraph, Telephone, Computer," *Social Research*, 64(3): 1067–1091, Fall 1997.

Oudeshorn, Nellie and Trevor Pinch (eds.). *How Users Matter; The Co-construction of Users and Technologies*. Cambridge, MA: The MIT Press, 2003.

Persson, Anders. "Intimacy among Strangers: On Mobile Telephone Calls in Public Places," *Journal of Mundane Behaviour*, 2(3): 1–7, October 2001.

Petrecca, Laura. "Cell Phone Marketers Calling All Pre-teens," *USA Today*, September 5, 2005.

Plant, Sadie. *On the Mobile: The Effects of Mobile Telephones on Social and Individual Life*. Motorolla, 2002. www.motorola.com/mot/doc/0/234_MotDoc.pdf (accessed June 5, 2006).

Pool, Ithiel de Sola (ed.). *The Social Impact of the Telephone*. Cambridge, MA: The MIT Press, 1977.

Pool, Ithiel de Sola. *Forecasting the Telephone; A Retrospective Technology Assessment*. Norwood, NJ: ABLEX Publishing Corporation, 1983.

Pool, Ithiel de Sola. "Will Mobile Telephones Move," in Tom Forrester (ed.), *The Information Technology Revolution*, pp. 144–145. Oxford, UK: Basil Blackwell Ltd., 1985.

Puttnam, David. *Bowling Alone: The Collapse and Revival of American Community*. New York: Simon and Schuster, 2000.

Rakow, Lana F. "Women and the Telephone: The Gendering of a Communications Technology," in Cheris Kramarae, (ed.), *Technology and Women's Voices: Keeping in Touch*. New York: Routledge and Kegan Paul, 1988.

Reinecke, Ian and Julianne Schultz. *The Phone Book: The Future of Australia's Communications on the Line*. Ringwood, NJ: Penguin Books, 1983.

Repacholi, Michael H. "WHO's EMF Project Results on RF Health Effects." http://www.who.int/emf/ (accessed 31 March, 2006).

Rhys-Morus, Iwan. "The Nervous System of Britain; Space, Time and the Electric Telegraph in the Victorian Age," *British Journal of the History of Science*, 33: 455–475, 2000.

Robbins, Kathleen A. and Martha A. Turner, "Chapter 6. United States: Popular Pragmatic and Problematic," in James E. Katz and Mark A. Aakhus (eds.), *Perpetual Contact: Mobile Communication, Private Talk, Public Performance*, pp. 80–93. Cambridge, UK: Cambridge University Press, 2002.

Schwartz-Cowan, Ruth. *A Social History of American Technology*. New York: Oxford University Press, 1997.

Silverstone, Roger and Eric Hirsch (eds.). *Consuming Technologies: Media and Information in Domestic Spaces*. London: Routledge, 1992.

Slesin, Louis. "WHO and Electrical Utilities: A Partnership on EMF's: Commentary: From the Field," *Microwave News*, October 1, 2005. http://www.microwavenews.com/fromthefield.html#partners (accessed March 31, 2006).

Smith, Merritt Roe. "Technological Determinism in American Culture," in Merritt Roe Smith and Leo Marx (eds.), *Does Technology Drive History? The Dilemma of Technological Determinism*, pp. 1–32. Cambridge, MA: The MIT Press, 1996.

Standage, Tom. *The Victorian Internet: The Remarkable Story of the Telegraph and the Nineteenth Century Online Pioneers*. London: Phoenix, 1998.

Steneck, Nicholas. *The Microwave Debate*. Cambridge, MA: The MIT Press, 1984.

Sterling, Bruce. "US Telephone Network," in Nick Heap, Ray Thomas, Geoff Einon, Robin Mason, and Hughie Mackay (eds.), *Information Technology and Society*, pp. 33–40. London: Sage, Open University, 1995.

Stern, Elle and Emily Gwathmey. *Once Upon a Telephone: An Illustrated Social History*. New York: Harcourt Brace and Company, 1994.

Stewart, William. *Independent Expert Group on Mobile Phones (IEGMP) Mobile Phones and Health*. UK: National Radiation Protection Board, 2000.

Toffler, Alvin. *The Third Wave*. London: Pan Books, 1981.

Trosby, Finn. "SMS, the Strange Duckling of GSM," *Telektronikk*, 187–194, March 2004.

Weed, Brad. "Visual Interaction Design: The Industrial Design of the Software Industry," *SIGCHI*, 28(3), July 1996.

Winner, Langdon. *The Whale and the Reactor*. Chicago, IL: University of Chicago Press, 1986.

Winston, Brian. *Media Technology and Society A History: From the Telegraph to the Internet*. London: Routledge, 1998.

Young, Peter. *Power of Speech: A History of Standard Telephones and Cables 1883-1983*. London: George Allen and Unwin, 1983.

Zimmerman Umble, Diane. "The Amish and the Telephone: Resistance and Reconstruction," Chapt. 11, in Roger Silverstone and Eric Hirsch (eds.), *Consuming Technologies: Media and Information in Domestic Spaces*. London: Routledge, 1992.

Index

About the Author

DAVID MERCER is Associate Professor in the Science Technology and Society Program at the University of Wollongong, Australia. He is a member of the editorial board of the journal *Metascience* and has published extensively in areas of social studies of science.